Power Plant Centrifugal Pumps
Problem Analysis and Troubleshooting

Power Plant Centrifugal Pumps
Problem Analysis and Troubleshooting

Maurice L. Adams Jr.

CRC Press
Taylor & Francis Group
Boca Raton London New York

CRC Press is an imprint of the
Taylor & Francis Group, an **informa** business

CRC Press
Taylor & Francis Group
6000 Broken Sound Parkway NW, Suite 300
Boca Raton, FL 33487-2742

© 2017 by Taylor & Francis Group, LLC
CRC Press is an imprint of Taylor & Francis Group, an Informa business

No claim to original U.S. Government works

Printed on acid-free paper
Version Date: 20161120

International Standard Book Number-13: 978-1-4398-1378-2 (Hardback)

This book contains information obtained from authentic and highly regarded sources. Reasonable efforts have been made to publish reliable data and information, but the author and publisher cannot assume responsibility for the validity of all materials or the consequences of their use. The authors and publishers have attempted to trace the copyright holders of all material reproduced in this publication and apologize to copyright holders if permission to publish in this form has not been obtained. If any copyright material has not been acknowledged please write and let us know so we may rectify in any future reprint.

Except as permitted under U.S. Copyright Law, no part of this book may be reprinted, reproduced, transmitted, or utilized in any form by any electronic, mechanical, or other means, now known or hereafter invented, including photocopying, microfilming, and recording, or in any information storage or retrieval system, without written permission from the publishers.

For permission to photocopy or use material electronically from this work, please access www.copyright.com (http://www.copyright.com/) or contact the Copyright Clearance Center, Inc. (CCC), 222 Rosewood Drive, Danvers, MA 01923, 978-750-8400. CCC is a not-for-profit organization that provides licenses and registration for a variety of users. For organizations that have been granted a photocopy license by the CCC, a separate system of payment has been arranged.

Trademark Notice: Product or corporate names may be trademarks or registered trademarks, and are used only for identification and explanation without intent to infringe.

Library of Congress Cataloging-in-Publication Data

Names: Adams, Maurice L., Jr. author.
Title: Power plant centrifugal pumps : problem analysis and troubleshooting / Maurice L. Adams, Jr.
Description: 2nd edition. | Boca Raton : Taylor & Francis, a CRC title, part of the Taylor & Francis imprint, a member of the Taylor & Francis Group, the academic division of T&F Informa, plc, [2016]
Identifiers: LCCN 2016044982| ISBN 9781439813782 (hardback) | ISBN 9781439813799 (ebook)
Subjects: LCSH: Hydroelectric power plants--Equipment and supplies--Maintenance and repair. | Centrifugal pumps--Maintenance and repair.
Classification: LCC TK1081 .A232 2016 | DDC 621.6/7--dc23
LC record available at https://lccn.loc.gov/2016044982

Visit the Taylor & Francis Web site at
http://www.taylorandfrancis.com

and the CRC Press Web site at
http://www.crcpress.com

Printed and bound in the United States of America by Sheridan

This book is dedicated to my late parents and late brother

Maury, Libby, and George

And to my late wives

Heidi and Kathy

And to my four mechanical engineering sons

Maury, Dr. Mike, RJ, and Nate

In Memoriam: Dr. Elemer Makay (1929–1996)—A Legend in Troubleshooting Power Plant Pumps

Elemer Makay was my closest friend for 27 years. He passed away in April 1996, ending his 5-year "troubleshooting effort" to prove the doctors wrong when they gave him 6 months to live after diagnosing his cancer in 1991. Just 2 weeks after Elemer received this bad news from the doctors, his younger of two adult daughters, Annie, was unexpectedly diagnosed with a terminal illness, and she passed away 2 weeks later. Elemer was the toughest guy I ever knew, and he surely was being tested beyond his full-rated capacity. Although he, of course, never got over the loss of Annie, Elemer did not give up, as would have been an easy out for one who was carrying his own terminal illness. He fought on for five more years. Even into the last months of his life, he continued his industry-recognized missionary efforts toward fixing all the electric utility industry's power plant pump problems. Shortly after the loss of his daughter Annie, he was blessed with a granddaughter, Katherine, by his older daughter, Virginia, and her husband, Jim, adding immeasurably to the pleasure and meaning of his last years.

Elemer was born in 1929 in Nyiregyhaza, Hungary. Preparing for a military career, he studied at the Nagy Kroly Military School from 1942 to 1945. After the war he continued his formal education at the University of Budapest. In 1948 he started his studies at the Hungarian Jozsef Nador University, majoring in mechanical engineering. He was an active freedom fighter in the 1956 Hungarian revolution. As the Soviet tanks suppressed that revolution, he escaped with many Hungarians over the frontier to Austria, traversing by foot the fields of not-yet-removed and still-active land mines that had remained buried from a prior Austro-Hungarian dispute. He later immigrated to the United States and entered graduate school at the University of Pennsylvania in Philadelphia while employed as a centrifugal pump designer and hydraulics specialist at a nearby major centrifugal pump manufacturer. He earned a PhD at Penn in mechanical engineering in 1966 with a research thesis in fluid dynamics. In 1969 he was employed by the Franklin Institute Research Laboratories (FIRL) in Philadelphia to start a new rotating machinery section.

I met Dr. Elemer Makay shortly after his arrival at FIRL. I was then in my second year of employment at FIRL, so we both had at least one thing in common—we were both recent "intruders" into this high-pressure dog-eat-dog organization in which individual senior staff survived mostly on the outside contract funds they could secure.

Because of common elements in our respective industrial experiences, we were drawn together by a sense of common interests. It quickly became obvious to us that we had far more in common. We soon became allies and solidified our deep friendship over the next 2 years. I left FIRL in July 1971 to work for Westinghouse's R&D Center near Pittsburgh.

A significant event occurred in April 1971. Northern States Power (NSP, Minneapolis), having recently read one of Elemer's early articles on feed water pump problems, contacted him for technical assistance in dealing with a quite nasty feed pump problem at its new Monticello Nuclear Power Plant. I strongly urged Elemer to act quickly on this great opportunity to be the utility company's expert against the pump vendor. Elemer's initial reluctance with my idea was because he had been that vendor's pump hydraulics specialist before coming to FIRL, and his departure from the vendor's employment was not on the warmest of terms with the management, to put it mildly.

It was easy for me to encourage him on this venture. I had nothing to fear of the pump vendor's reaction if Elemer was involved on NSP's behalf. He knew quite well why the pumps sold to NSP were not well suited to the application, which was basic information the pump vendor was apparently not sharing with NSP. After a long lunch to discuss NSP's request, Elemer and I returned to our respective offices at FIRL. Elemer was still undecided what to tell Roland Jensen of NSP, who later became senior vice president of power production, and Dave McElroy of NSP, who later became chairman and president of NSP.

Not long after that lunch, our secretary was searching various offices for Elemer to take a "very important long-distance call from some company in Minnesota." I told her to transfer the call to my phone, at which time I "dealt myself in" by telling Mr. Jensen that Elemer and I would make their 7:00 p.m. urgent meeting that day at NSP Corporate Offices. I then went quickly to our travel person and signed out for two air tickets to Minneapolis and two hundred dollars cash for each of us. I waited briefly for Elemer back in his office, handed him his ticket and the money, and informed him that my wife, Heidi, was meeting us at the Philadelphia airport (not far from where I lived) with some changes of clothing. For a few moments Elemer was speechless, a rare event for him. This was Elemer's first outing as what he was eventually to be known as: the world's leading troubleshooter for power plant pumps. Over the next 25 years, a year never went by without Elemer and I (two natural-born renegades who enjoyed rocking the boat) jovially reliving and savoring that NSP day many times over, each time enjoying it a little more than the previous time.

In 1973 Elemer started his own company and went on to single-handedly troubleshoot and correct the most troublesome pump problems that had plagued power plants nationwide for many years, in particular boiler and nuclear feed water pump problems. Naturally, when a single individual successfully takes a whole industry to task for its shortcomings, that individual

In Memoriam

will accumulate some detractors. Elemer's detractors were primarily the management groups in the pump companies. But those individuals faded into obscurity while Elemer became a legend in the electric utility industry.

Based on Elemer's persuasion and guidance, the Electric Power Research Institute (EPRI) launched a ten million dollar multiyear project in the mid-1980s to fund badly needed research for the improvement of boiler feed water pump design technology. Elemer and I were the two principals on the EPRI consultant's team that awarded and steered that EPRI research project. It involved nearly all the most capable centrifugal pump technologists from both sides of the Atlantic Ocean and significantly advanced the engineering science of high-energy centrifugal pumps.

Elemer's articles and papers are studied and applied by pump designers and researchers worldwide. He often credited his success to his having the "largest laboratory in the world, from the Atlantic to the Pacific," namely, the entire U.S. electric utility industry. In 1992 Elemer was awarded the highest honor for a pump technologist: the ASME Worthington Award. Elemer has left his mark on his profession to an extent that few individuals ever achieve.

Many times Elemer and I discussed the need for a book on troubleshooting power plant pumps. But in the midst of our busy lives, time ran out for Elemer. This book clearly reflects the rich record of Elemer's power plant pump troubleshooting successes documented in his many lucid published papers, articles, reports, and short-course handouts.

Contents

In Memoriam: Dr. Elemer Makay (1929–1996)—A Legend
in Troubleshooting Power Plant Pumps vii
Preface xv
Acknowledgments xvii
Author xix

Section I Primer on Centrifugal Pumps

1. Pump Fluid Mechanics, Concepts, and Examples 3
   1.1   Flow Complexity and Flow-Path Geometry 3
   1.2   Centrifugal Pump Theory 10
        1.2.1   Pump Impeller Flow and Head Production 10
        1.2.2   Specific Speed and Design Parameters 15
   1.3   Centrifugal Pump Configurations 16
        1.3.1   Pump Casing Entrance and Discharge Flow Paths 16
        1.3.2   External and Internal Return Channels of Multistage
                Pumps 20
        1.3.3   Pump Priming 21
        1.3.4   Controls 22

2. Pump Performance, Terminology, and Components 23
   2.1   Hydraulic Performance and Efficiencies 23
   2.2   Intersection of Pump and System Head–Capacity Curves 24
   2.3   Cavitation Damage and Pump Inlet Suction-Head
        Requirements 25
        2.3.1   Description of Pump Cavitation Phenomenon 25
        2.3.2   Laboratory Shop Testing to Quantify Pump
                Cavitation Incipience 27
        2.3.3   Required Net Positive Suction Head and Available
                Net Positive Suction Head 28
        2.3.4   Operation of Pumps in Parallel 29
   2.4   Mechanical Components 30
        2.4.1   Shafts 30
        2.4.2   Couplings 30
        2.4.3   Bearings 35
        2.4.4   Seals 39
        2.4.5   Thrust Balancers 45
   2.5   Drivers 47

3. Operating Failure Contributors 49
3.1 Hydraulic Instability and Pressure Pulsations 49
3.1.1 Head–Capacity Curve Instability 49
3.1.2 Pressure Pulsation Origins 51
3.1.3 Criteria for Minimum Recirculation Flow 52
3.2 Excessive Vibration 53
3.2.1 Rotor Dynamical Natural Frequencies and Critical Speeds 55
3.2.2 Self-Excited Dynamic-Instability Rotor Vibrations 57
3.2.3 Dynamic Forces Acting on the Rotor 59
3.3 Wear 60
3.3.1 Damage Caused by Pump Cavitation 61
3.3.2 Adhesive Wear 62
3.3.3 Abrasive Wear 64
3.3.4 Delamination Wear 64
3.4 Operating Problem Modes 64
3.4.1 Rotor Mass Unbalance 65
3.4.2 Unfavorable Rotor Dynamic Characteristics 66
3.4.3 Hydraulic Forces at Off-Design Operating Conditions 66
3.4.4 Dynamic Characteristics of Foundation, Support Structure, and Piping 69
3.4.5 Unfavorable Pump Inlet Flow Conditions 70
3.4.6 Bearing, Seal, Shaft, and Thrust Balancer Damage 70
3.5 Condition Monitoring and Diagnostics 72
3.5.1 Vibration Measurement 73
3.5.2 Pressure Pulsation Measurement 76
3.5.3 Temperature Measurement 77
3.5.4 Cavitation Noise Measurement 78
3.5.5 Pump Test Rig for Model-Based Condition Monitoring 78
3.5.6 Summary of Monitoring and Diagnostics 82

Section II Power Plant Centrifugal Pump Applications

4. Pumping in Fossil Plants 85
4.1 Boiler Feed Water 85
4.2 Condensate, Heater Drain, and Condenser Circulating 89
4.3 Boiler Circulating 93

5. Pumping in Nuclear Plants 95
5.1 Pressurized Water Reactor Primary Reactor Coolant 98
5.2 Feed Water and Auxiliary Feed Water 107
5.3 Residual Decay Heat Removal 107
5.4 High Pressure Safety Injection and Charging 107

Contents xiii

  5.5 Boiling Water Reactor Main Circulating 108
  5.6 Quarterly Testing of Standby Safety Pumps................................... 109

Section III Troubleshooting Case Studies

6. Boiler Feed Pump Rotor Unbalance and Critical Speeds 113
  6.1 Case 1 114
  6.2 Case 2 117
  6.3 Case 3 119
  6.4 Summary... 122

7. Nuclear Feed Pump Cyclic Thermal Rotor Bow 123
  7.1 Background on Cyclic Vibration Symptom................................... 123
  7.2 Rotor Vibration Analyses... 123
  7.3 Cyclic Thermal Bow Analysis 125
  7.4 Shop Cyclic Thermal Test and Low-Cost Fix 125

8. Boiler Circulation Pump 127
  8.1 Problem Background 127
  8.2 Investigation 129
  8.3 Floating-Ring Seal Leakage 130
  8.4 Wear Ring Leakage 131
  8.5 Root Cause and Fixes.. 131

9. Nuclear Plant Cooling Tower Circulating Pump 133
  9.1 Problem Background 133
  9.2 Investigation and Root Cause 133
  9.3 Fixes 135

10. Condensate Booster Pump Shaft Bending 137
  10.1 Problem Background 137
  10.2 Investigation 137
  10.3 Root Cause and Fix 141

11. Pressurized Water Reactor Primary Coolant and Auxiliary Feed Pumps.. 143
  11.1 Primary Coolant Pump (PCP) Problem Background 143
  11.2 Vibration Instrumentation False Alarm 143
  11.3 Worn Impeller Journal Bearing... 144
  11.4 Primary Coolant Pump Summary 144
  11.5 Auxiliary Feed Pump Problem Background 145
  11.6 Auxiliary Feed Pump Analysis Results and Recommendations ... 145

12. Cases from Mechanical Solutions, Inc. 149
    12.1   Below-Ground Resonance of Vertical Turbine Pump 149
    12.2   Cavitation Surge in Large Double-Suction Pumps 150
    12.3   Impeller Vane-Pass Excitation of a Pipe Resonance
           in a Nuclear Plant 153

Bibliography 155

Index 157

Preface

This book is comprised of three sections that provide complete background and breadth of material to meet the needs of newcomers to the field as well as experienced specialists. Its focus is on the centrifugal type pump since that is the dominant pump type employed in power plants. Other industries heavily reliant upon centrifugal pumps (e.g., ocean ship propulsion systems, petrochemical process plants) can also benefit from this book.

Section I "Primer on Centrifugal Pumps" is a group of three chapters treating centrifugal pump fundamentals (Chapter 1), performance and components (Chapter 2), and operating failures (Chapter 3).

Section II "Power Plant Centrifugal Pump Applications" is composed of two chapters describing pumping in fossil plants (Chapter 4) and pumping in nuclear plants (Chapter 5).

Section III "Troubleshooting Case Studies" is seven chapters detailing actual troubleshooting case studies for several power plant pump problems. These case studies are taken both from the author's own pump troubleshooting projects as well as documented case studies from articles, conference papers, short courses, and private communications with the late Dr. Elemer Makay. Case study documentaries have also been provided by Mechanical Solutions, Inc.

Acknowledgments

Truly qualified technologists invariably acknowledge the shoulders upon which they stand. I am unusually fortunate in having worked for several expert caliber individuals during my formative 14 years of industrial employment prior to becoming a professor in 1977, especially my 4 years at the Franklin Institute followed by my 6 years at the Westinghouse Corporate R&D Center's Mechanics Department. I am also highly appreciative of many subsequent rich interactions with other technologists. I here acknowledge those individuals, many of whom have unfortunately passed away over the years. They were members of a now-extinct breed of giants who unfortunately have not been replicated in today's industrial workplace environment.

My work in centrifugal pumps began in the mid-1960s at Worthington Corporation's Advanced Products Division (APD) in Harrison, New Jersey. There I worked under two highly capable European-bred engineers, chief engineer Walter K. Jekat (German) and his assistant John P. Naegeli (Swiss). John Naegeli later returned to Switzerland and eventually became general manager of Sulzer's Turbo-Compressor Division and later general manager of its Pump Division. The APD general manager was Igor Karassik, the world's most prolific writer of centrifugal pump articles, papers, and books, and an energetic teacher on centrifugal pumps for all the recent engineering graduates at APD like me. My first assignment at APD was basically to be "thrown into the deep end" of a new turbomachinery development for the U.S. Navy that even today would be considered highly challenging. That new product was comprised of a 42,000 rpm rotor having an overhung centrifugal air compressor impeller at one end and an overhung single-stage impulse steam turbine powering the rotor from the other end, with water-lubricated turbulent fluid film bearings. Worthington sold several of these units over a period of many years.

In 1967 I seized on an opportunity to work for an internationally recognized group at the Franklin Institute Research Laboratories (FIRL) in Philadelphia. I am eternally indebted to several FIRL technologists for the knowledge I gained from them and for their encouragement for me to pursue graduate studies part time, which led to my engineering master's degree from a Penn State extension near Philadelphia. The list of individuals I worked under at FIRL is almost a who's who list for the field and includes the following: Elemer Makay (centrifugal pumps), Harry Rippel (fluid-film bearings), John Rumbarger (rolling-element bearings), and Wilbur Shapiro (fluid-film bearings, seals, and rotor dynamics). I also had the privilege of working with a distinguished group of FIRL's consultants from Columbia University, specifically, Professors Dudley D. Fuller, Harold G. Elrod, and Victorio "Reno" Castelli.

My Franklin Institute job gave me the opportunity to publish in my field. That national recognition helped provide my next job opportunity when, in 1971, I accepted a job in what was truly a distinguished industrial research group, the Mechanics Department at Westinghouse's R&D Center near Pittsburgh. The main attraction of that job for me was my new boss, Dr. Albert A. Raimondi, leader of the bearing mechanics section, whose famous papers on fluid-film bearings I had been using ever since my days at Worthington. An added bonus was the presence of the person holding the department manager position: A. C. "Art" Hagg, the company's internationally recognized rotor vibration specialist. My many interactions with Art Hagg were all professionally enriching. At Westinghouse I was given the lead role on several cutting-edge projects, including nonlinear dynamics of flexible multibearing rotors for large steam turbines and reactor coolant pumps, bearing load determination for vertical multibearing pump rotors, seal development for refrigeration centrifugal compressors, and turning-gear slow-roll operation of journal bearings, developing both test rigs and new computer codes for these projects. I became the junior member of an elite ad hoc trio that included Al Raimondi and D. V. "Kirk" Wright (manager of the dynamics section). They encouraged me to pursue a PhD part time, which I completed at the University of Pittsburgh in early 1977. I express special gratitude to my PhD thesis advisor at Pitt, Professor Andras Szeri, who considerably deepened my understanding of the overlapping topics of fluid dynamics and continuum mechanics.

Since entering academia in 1977 I have benefited from the freedom to publish widely and to apply and extend my accrued experience and knowledge through numerous consulting projects for turbomachinery manufacturers and electric utility companies. I appreciate the many years of support for my funded research provided by the Electric Power Research Institute (EPRI) and the NASA Glenn Research Laboratories. Academic freedom has also made possible opportunities to work abroad with some highly capable European technologists, specifically at the Brown Boveri Company (BBC, now within Alstom Power), Sulzer Co. (Winterthur, Switzerland), KSB Pump Company (Frankenthal, Germany), and the Swiss Federal Institute (ETH, Zurich). At BBC (Baden, Switzerland) I developed a lasting friendship with my host, Dr. Raimund Wohlrab. At the Sulzer Pump Division I was fortunate to interact with Dr. Dusan Florjancic (engineering director), Dr. Ulrich Bolleter (vibration engineering), and Dr. Johan Guelich (hydraulics engineering). At the KSB Pump Company I was fortunate to interact with Peter Hergt (head of KSB's Central Hydraulic R&D, 1975–1988) and his colleagues. I particularly value the lasting friendship developed with my host at the ETH, the late Dr. Georg Gyarmathy, the ETH turbomachinery professor, 1984–1999. This book rests upon the shoulders of all whom I have here acknowledged.

Author

Maurice L. Adams Jr. is founder and past president of Machinery Vibration Inc. and is professor emeritus of mechanical and aerospace engineering at Case Western Reserve University. The author of more than 100 publications and the holder of U.S. patents, he is a life member of the American Society of Mechanical Engineers. Adams earned a BSME (1963) at Lehigh University, an MEngSc (1970) at the Pennsylvania State University, and a PhD (1977) in mechanical engineering at the University of Pittsburgh. Adams worked on rotating machinery engineering for 14 years in industry prior to becoming a professor in 1977, including employment at Allis Chalmers, Worthington, Franklin Institute Research Laboratories, and Westinghouse Corporate R&D Center. He was the recipient in 2013 of the Vibration Institute's Jack Frarey Medal for his contributions to the field of rotor dynamics.

Section I

Primer on Centrifugal Pumps

1
Pump Fluid Mechanics, Concepts, and Examples

Centrifugal pumps are used in the majority of all fluid flow processes. The complexities of the fluid flow patterns and design practicalities for these pumps have involved intensive engineering endeavors for well over 100 years. The requirements of these pumps vary considerably for different industries and applications. Power plant requirements for centrifugal pumps are among the most demanding of any industry. With the advent of modern computational fluid mechanics (CFM), finite element analyses (FEA), measurement sensors, and digital data processing, the evolution of centrifugal pump technology development still continues to advance.

1.1 Flow Complexity and Flow-Path Geometry

The flows in both stationary and rotating internal flow passages of centrifugal pumps are quite complicated. Even when operating at the best efficiency point (BEP) 100% design flow, centrifugal pump internal flow fields are somewhat unsteady. At off-design operating flows the internal flows are highly unsteady. Showing still pictures like Figure 1.1 does not convey this flow unsteadiness, providing only instantaneous snap shots of typical flow fields within and around the rotating impeller flow passages.

Centrifugal pumps are closely related to hydro turbines, the fundamental distinction being that in turbines mechanical energy is extracted from the fluid, whereas in pumps mechanical energy is transferred to the fluid. This distinction results in a major fluid dynamical difference between centrifugal pumps and hydro turbines, namely that within turbine impellers the fluid is accelerated like in a *nozzle*, while within pump impellers the fluid is decelerated like in a *diffuser*. This fundamental difference makes highly efficient hydraulic design considerably more challenging for centrifugal pumps than for turbines. An insightful way to understand this fundamental difference is to view the pump impeller as an assembly of rotating diffuser channels and to view the turbine impeller as an assembly of rotating nozzles. Figure 1.2 shows a clear visual distinction between centrifugal pump and hydro turbine impellers. One immediately notices that a pump impeller typically has

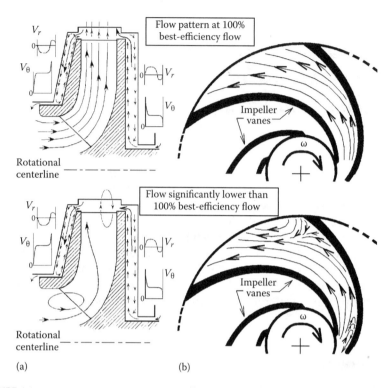

FIGURE 1.1
Centrifugal pump impeller snapshots of flow patterns. (a) Radial flow views and (b) circumferential flow views relative to rotating impeller.

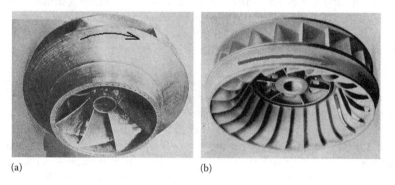

FIGURE 1.2
Comparison between Francis impellers for (a) a pump and (b) a turbine.

about one-third as many vanes as a turbine impeller. One also immediately notices that a pump impeller vane correspondingly wraps around the impeller about three times the wrap angle of a turbine vane.

Figure 1.3 illustrates the fundamental fluid dynamics difference between a diffuser and a nozzle. In the direction of flow a nozzle's flow area decreases

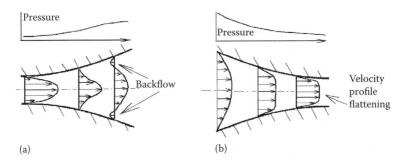

FIGURE 1.3
Velocity profiles and pressures for (a) diffusers and (b) nozzles.

and thus average fluid velocity increases, and in consequence it yields static pressure decreasing and velocity profile flattening in the direction of flow. In contrast, a diffuser's flow area increases and thus average fluid velocity decreases in the direction of flow, and in consequence it yields static pressure increasing and velocity profile arching in the direction of flow. When the degree of flow deceleration in a diffuser exceeds some critical level, the adverse pressure gradient combined with velocity profile arching initiates backflow (called *flow separation*) at the flow channel boundary. Flow separation is a churning action that significantly wastes flow energy through viscous dissipation into heat. Clearly, flow separation is not consistent with high-efficiency hydraulic performance. An efficient diffuser must be long enough so that its flow area increase rate is sufficiently gentle to avoid flow separation. In contrast, an efficient nozzle can be relatively short since flow separation is not a concern, given a nozzle's inherent favorable pressure gradient and velocity profile flattening. Understanding this basic contrast between a diffuser and a nozzle leads one to a basic insight into why highly efficient hydraulic design is more challenging for centrifugal pumps than for turbines.

As typified in Figure 1.2, the contrasting differences between pump impeller geometry and turbine impeller geometry clearly reflect this fundamental difference between diffusers and nozzles. That is, the relatively long wrapped-around pump impeller vanes form a considerably more gentle flow area transition than the relatively shorter turbine vanes. Optimum turbine impeller geometry benefits from shorter vanes by allowing many more vanes for better flow guidance without excessive eye exit flow blockage.

Realizing the basic differences between nozzles and diffusers, one can also appreciate the well-known fact that a typical-high efficiency centrifugal pump impeller can be a fairly efficient turbine impeller. Conversely, a typical high efficiency turbine impeller will make a quite inefficient centrifugal pump impeller. Accordingly, impellers in so-called pump-turbines are in reality slightly modified pump impellers, which are highly efficient both in the pumping and the turbine operating modes. As well known, the role of a pump-turbine is to pump water from a lower-elevation water source to a higher-elevation reservoir during the lower-demand hours of the day, and then to reverse the water flow for the turbine mode to generate electricity during the high-demand hours. Such facilities are often called pump-storage units since energy is stored by the water pumped into the higher-elevation reservoir.

The impeller in a centrifugal pump is where the mechanical energy is transferred into fluid dynamical energy, and thus the geometric configuration of a centrifugal pump impeller is the most important property of the pump affecting energy efficiency. However, the geometric configurations of the stationary nonrotating flow channels directing the flow into and out of the impeller, collectively called the *pump casing*, are also important features of the pump affecting energy efficiency. Since the pump casing does not add any energy to the pumped fluid, energy considerations in casing design are focused exclusively on minimizing energy losses. In addition, casing flow-path geometry is also important for avoidance of cavitation and excessive flow-induced vibration.

The ideal inlet channel flow of a centrifugal pump suction nozzle is a nearly uniform fluid velocity profile into the suction side eye of the impeller, for example, as illustrated by the nozzle flow velocity profiles in Figure 1.3. A straight section of nozzle pipe several diameters long with a gentle flow area contraction is the type of pump suction nozzle that yields a nearly uniform inlet velocity distribution. However, because of constraints often imposed upon pump placement and piping arrangements, a long and straight suction nozzle is frequently not feasible. Furthermore, in multistage and double-suction pumps it is simply not geometrically possible for a single long and straight suction nozzle to feed all impeller inlets. Figure 1.4 illustrates the simplest and most efficient inlet nozzle configurations. In applications where

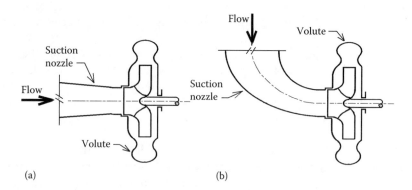

FIGURE 1.4
Optimum inlet nozzles for single-impeller single-suction pumps. (a) Straight suction nozzle and (b) gentle-elbow suction nozzle.

piping arrangements and/or radial bearing placement do not allow using either of the Figure 1.4 suction nozzles, the *volute suction nozzle* illustrated in Figure 1.5 is often used, such as for the first stage of multistage pumps where the shaft must penetrate through the first stage.

The multistage pumps illustrated in Figure 1.6 typify the use of a suction volute inlet to the first stage of a multistage pump. Impeller discharge and suction flow paths downstream of the first-stage impeller do not have very much space to diffuse impeller discharge flow and guide the flow into the suction inlet of the next impeller or the pump final discharge chamber. Development of efficient multistage centrifugal pumps therefore has required extensive laboratory performance testing of various configurations of pump internal casing flow path channel geometry.

Having sufficiently low inlet flow velocities at the impeller eye to avoid cavitation, the optimum hydraulic configuration is often the *double suction* casing design. The double-suction design is essentially a configuration with an impeller that resembles two single-suction impellers mounted back-to-back so that the impeller has two inlets and one discharge. This approach is common in boiler feed water booster pumps and in nuclear feed water pumps to avoid cavitation. Some multistage feed pumps have a double-suction first stage as an alternative to having a separate booster pump. In contrast to fossil fuel-fired power plants, feed water pumps for nuclear power plants are typically single-stage double-suction pumps as typified by the nuclear feed pump illustrated in Figure 1.7. An observant walk through the typical power plant reveals that many different types of pump services employ a double-suction centrifugal pump configuration, the design motivation usually being that a double-suction design is often the optimum approach to avoid cavitation. Cavitation is an extremely important topic in centrifugal pump design and operation, and is separately treated later in this chapter.

The energy imparted to a fluid by a centrifugal pump impeller is contained in two parts: (1) the fluid's pressure increase through the impeller

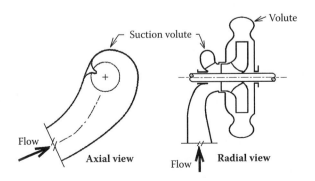

FIGURE 1.5
Suction volute inlet flow channel.

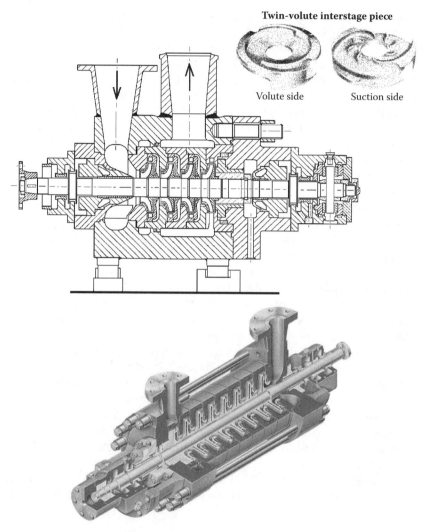

FIGURE 1.6
Multistage pumps.

(the major portion), plus (2) the fluid's velocity increase through the impeller (i.e., kinetic energy). The velocity of fluid just as it leaves the impeller is generally much higher than fluid velocity exiting at the pump discharge flange. That is, the pump discharge fluid average velocity essentially matches the average fluid velocity in the piping, which transports the pumped fluid away from the pump to its destination. Clearly, efficient hydraulic design needs smooth flow deceleration/diffusion to maximize the amount of available fluid kinetic energy at the impeller exit that is transformed into

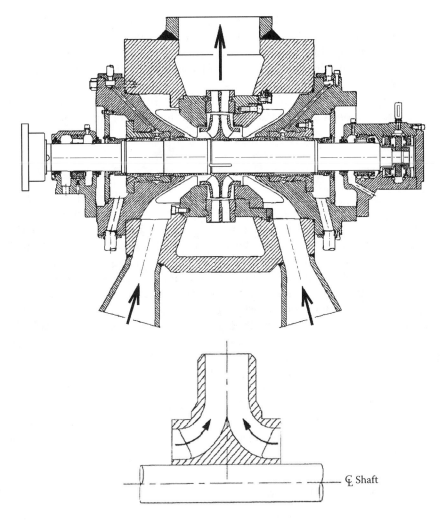

FIGURE 1.7
Typical double-suction feed pump for nuclear power plants.

additional pump discharge pressure energy à la Bernoulli's equation, as follows:

$$h + \frac{p}{\gamma} + \frac{v^2}{2g} = \text{Energy. For zero elevation change } (\Delta h = 0)$$

and 100% efficient diffusion yields $\Delta p = \frac{\gamma v^2}{2g}$

As illustrated in Figure 1.3 and previously explained in comparing centrifugal pump impellers to turbine impellers, more care is required to efficiently decelerate flow than to efficiently accelerate flow. Therefore, configuring the casing discharge section for energy efficient collection and diffusion of impeller discharge flow requires an even more challenging hydraulic design endeavor than for the suction section of the casing. The high power consumption of large centrifugal pumps that permeate electric power generating plants places a much higher priority on pump hydraulic efficiency than in many other industrial applications using smaller power centrifugal pumps.

1.2 Centrifugal Pump Theory

Centrifugal pumps were already a well-developed mature type of machinery before the advent of the modern high-speed digital computer. Today's pump designers, of course, have available general-purpose computer codes for developing the impeller and casing geometry for any new design. But as is now widely appreciated by savvy engineers, many users of modern engineering computer codes (most prominently finite element analysis, FEA) are clueless on the fundamental theory and limitations of what is inside the code. A basic understanding of how a centrifugal pump works best starts with the elementary precomputer explanation of how the pressure rise and flow of a centrifugal pump are produced. One of the early books on centrifugal pumps most frequently referenced in its time is that of Pfleiderer (1932), which is in German. The book in English most frequently referenced is by Stepanoff (1957). But the published literature on the subject is now enormous. The presentation here on the basic theory of centrifugal pumps is mostly inspired by Stepanoff.

1.2.1 Pump Impeller Flow and Head Production

The elementary theory starts with what are commonly called the *velocity triangles*. They are constructed from the three velocity vectors at a particular point in the pump impeller flow path as follows: (1) absolute impeller peripheral (tangential) velocity, u; (2) fluid velocity relative to impeller, w; and (3) absolute fluid velocity, c. While such velocity vector triangles can be made to represent any flow point within the impeller, focus is usually upon the impeller *flow entrance* (subscripted "1") and impeller *flow exit* (subscripted "2"). Figure 1.8 shows a typical looking set of entrance and exit velocity vector triangles. The angles α and β shown on the velocity triangles are for the fluid velocity angle relative to the impeller (β) and the absolute fluid velocity angle (α). The illustrated vertical velocity components c_{m1} and c_{m2} are the absolute radial velocity components and thus correspond to the impeller net through flow. Here all the velocity vectors are taken as the average velocities over the plane normal to the average flow relative to the impeller at

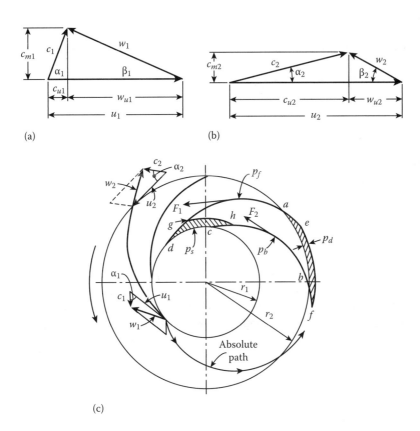

FIGURE 1.8
Example impeller velocity triangles at entrance and discharge points. (a) Entrance, (b) discharge, and (c) superimposed on impeller vanes with impeller forces.

that point. This is, of course, an engineering approximation of the actual 3D velocity distribution. Modern computer codes for modeling impeller flow fields are not limited by this approximation. However, this traditional pre-computer design-theory approach still provides a clear picture of how the pump impeller produces *head*. That is, utilizing Newton's second law for rotational motion, that is, the time rate of change of angular momentum vector $\vec{h}$ is equal to the applied torque vector $\vec{T}$ causing that angular momentum change. The $\vec{h}$ and $\vec{T}$ vectors align with the axis of impeller rotation, so their scalar axial magnitudes T and h can be used here.

$$T = \frac{dh}{dt} \tag{1.1}$$

The impeller net driving power P (excluding various friction losses) is then given by Equation 1.2, where $dm/dt = Q\gamma/g$ is the net mass flow rate passing through the impeller, Q is the volume flow rate, γ is the fluid weight density, g is the gravitational constant, and ω is the impeller angular velocity:

$$P = T\omega = \frac{Q\gamma}{g}\omega(r_2 c_2 \cos\alpha_2 - r_1 c_1 \cos\alpha_1) \qquad (1.2)$$

Substituting $u = \omega r$ into Equation 1.2 condenses the equation for the net impeller input power P applied to the liquid by the impeller vanes to the following equation:

$$P = \frac{Q\gamma}{g}(u_2 c_{u2} - u_1 c_{u1}) \qquad (1.3)$$

Neglecting for the time being the energy losses between impeller output and the hydraulic power determined from external flow and pressure measurements, the following expression results by equating measured hydraulic power output of a pump stage to the impeller power applied to the liquid (Equation 1.3).

$$Q\gamma H_i = \frac{Q\gamma}{g}(u_2 c_{u2} - u_1 c_{u1}) \qquad (1.4)$$

Eliminating $Q\gamma$ from Equation 1.4 yields Euler's equation for H_i, the *input head* (units of length), as follows:

$$H_i = \frac{u_2 c_{u2} - u_1 c_{u1}}{g} \qquad (1.5)$$

For a pump stage with assumed zero-energy input head at the impeller suction inlet point, Equation 1.5 simplifies Euler's equation as follows:

$$H_i = \frac{u_2 c_{u2}}{g} \qquad (1.6)$$

Applying the trigonometry law of cosines to the velocity vector triangles in Figure 1.8 transforms Equation 1.4 as follows:

$$H_i = \frac{(c_2^2 - c_1^2) + (u_2^2 - u_1^2) + (w_1^2 - w_2^2)}{2g} \qquad (1.7)$$

Pump Fluid Mechanics, Concepts, and Examples

$\left(c_2^2 - c_1^2\right)/2g$ gives the kinetic energy (KE) portion of H_i, while $\left[\left(u_2^2 - u_1^2\right) + \left(w_1^2 - w_2^2\right)\right]/2g$ is the pressure-increase portion of H_i.

In an oversimplified pseudo 1D concept of flow through an impeller, one would visualize the flow streamlines relative to the impeller as closely conforming to the flow channel shape between two adjacent vanes (e.g., see Figure 1.1 for the 100% design flow case). However, the actual velocity distribution within the impeller is far more complicated for a number of fundamental reasons.

First, in order for an impeller vane to apply energy to the liquid, the vane surface that is "pushing" the liquid must have a higher average pressure distribution upon it than the adjacent vane surface that is "pulling" the liquid. So each vane has a high-pressure side and a low-pressure side consistent with the impeller continuously adding energy to the liquid as it progresses through the impeller from entrance to discharge.

Second, the liquid's rotary inertia between two adjacent vanes will resist rotating with the impeller's speed. So a counterrotational flow recirculation relative to the impeller will occur between each set of adjacent vanes, as illustrated in Figure 1.9. This relative recirculation flow is superimposed upon the through flow, thus tending to produce lower relative velocity nearer the high-pressure side of a vane and a higher relative velocity nearer the low-pressure side of a vane.

Third, pumps frequently are operated over a broad range of the usable flow capacity of the pump. The patterns and unsteadiness of impeller flow become more pronounced at off-design capacities than at the BEP 100% flow condition, thus further complicating the actual impeller 3D flow patterns.

Clearly, the precomputer age development of centrifugal and axial flow pump designs necessitated considerable laboratory testing of prospective pump flow-path configurations. Accordingly, such test results have traditionally been highly proprietary information within each of the major pump

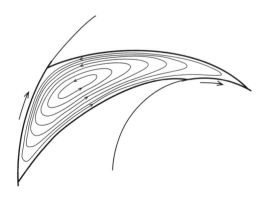

FIGURE 1.9
Relative circulation between two adjacent vanes.

manufactures. While present computation fluid mechanics (CFM)-based pump design modeling computer codes have surely reduced the degree of laboratory testing needed, research and development tests have remained essential. For example, to provide the CFD codes with various empirical input phenomenological coefficients (e.g., viscosity, density, vapor pressure versus temperature, etc.), testing is required to hone the CFD codes' inputs for optimum accuracy in predicting pump performance; that is, testing to "calibrate" the CFD codes sufficiently to eliminate or at least significantly reduce the need for performance testing. This collaborative interaction between CFD code use and empirical-based code inputs permeates all present turbomachinery development, including modern aircraft gas-turbine jet engines.

To avoid confusing the input head (H_i) and the Euler head (H_e) will be used here. For the simplest Euler expression (no entrance energy, e.g., zero prerotation of impeller entrance flow) the following expression is obtained from the discharge velocity triangle and Equation 1.6:

$$H_e = \frac{u_2^2}{g} - \frac{u_2 c_{m2}}{\tan \beta_2} \qquad (1.8)$$

This provides an equation of a straight line essentially showing head versus flow for $\beta_2 < 0$ for 100% efficiency as illustrated in Figure 1.10. For efficient pumps, the normal range for vane discharge angle is $20° < \beta_2 < 25°$. Due to the recirculation of liquid within the impeller, as illustrated in Figure 1.9, the actual discharge angle of the liquid relative to the impeller (say β_2') will be smaller than the vane discharge angle β_2. And the actual entrance angle of the liquid relative to the impeller (β_1') will be larger than the vane entrance angle β_1. Altering the velocity triangles in Figure 1.8 to reflect the actual liquid entrance and discharge angles relative to the impeller shows why the

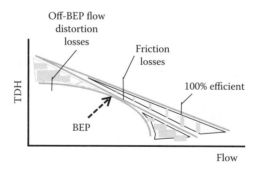

FIGURE 1.10
Total dynamic head (TDH) versus flow; zero entrance energy, $\beta_2 < 90°$.

actual energy absorbed by the liquid will be less than that given by Euler's equation.

1.2.2 Specific Speed and Design Parameters

There are several so-called *dimensionless numbers* that characterize fluid dynamics, heat transfer, and other engineering specialties. Some of the most commonly recognized ones include the Reynolds, Mach, Weber, Froude, Nusselt, and Prandtl numbers. The most relevant number for a centrifugal pump stage is *specific speed*. Although it is expressible as a dimensionless number, it is not dimensionless when a mix of system units is employed, as is common for centrifugal pumps. A dimensionless specific speed in English or SI units can be expressed as follows:

$$n_s = \frac{nQ^{1/2}}{(gH)^{3/4}}$$

where n = speed (rev/sec), Q = flow (ft³/sec), H = head (ft), and g (ft/sec²); or in SI units where n (rev/sec), Q (m³/sec), H (m), and g (m/sec²).

However, the specific speed expression most commonly used in the United States employs the following mix of units, which is not dimensionless:

$$N_s = \frac{n\sqrt{Q}}{H^{3/4}} \tag{1.9}$$

where the units used are Q (gpm), H (ft), and n (rpm). The specific speed used in Europe is the same expression as Equation 1.9 employing SI units, but likewise is not dimensionless, as follows.

In the now widely referenced highly informative experienced-based Worthington illustration shown in Figure 1.11, one immediately sees that independent of impeller size, the optimum-homologous shapes of the shown impeller geometries evolved uniquely as a function of specific speed, as are the shown characteristic performance curves for head (H), efficiency (E), and power (P) versus flow (Q). Also, the family of efficiency curves in Figure 1.11 shows the effect of pump size on efficiency, reflecting that overall efficiency losses do not scale up in proportion to pump size. Additionally, the performance curves illustrated at the top of Figure 1.11 show that for high-specific-speed pumps, the required driving power P maxes as the flow is reduced. Whereas for low-specific-speed pumps, the power maxes as flow is increased. The most important observation from this is that for high-specific-speed pumps where the drive motor is sized for a prescribed flow range around it best efficiency point, the motor will become overloaded at

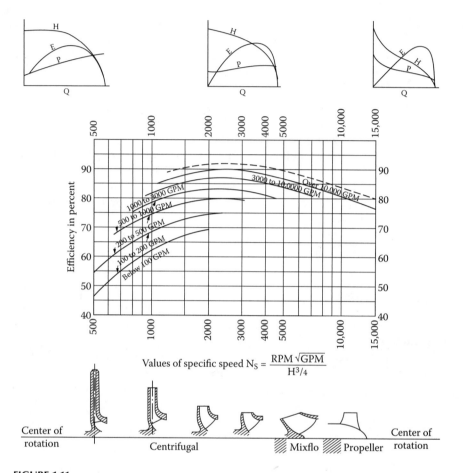

FIGURE 1.11
Efficiency (E), head (H), and power (P) versus Flow (Q) and optimum impeller shapes as functions of specific speed (N_S).

low flows. Therefore never start a high-specific-speed pump at the closed-valve shut-off condition. Similarly for the same reason, to avoid driver overload, do not start a low-specific-speed pump at the valve wide-open runout flow condition.

1.3 Centrifugal Pump Configurations

1.3.1 Pump Casing Entrance and Discharge Flow Paths

The impeller entrance/inlet flow and discharge flow paths are important considerations in a well-designed centrifugal pump. The most desirable inlet flow condition is one with nearly uniform axial flow velocity, the objective being a smooth efficient entrance into the impeller. The most desirable discharge flow condition is one that maximally recaptures pressure energy from the impeller discharge flow kinetic energy (velocity head), previously shown in Section 1.1. with the Bernoulli equation:

$$h + \frac{p}{\gamma} + \frac{v^2}{2g} = \text{Energy. For zero elevation change } (\Delta h = 0)$$

$$\text{and 100\% efficient diffusion yields } \Delta p = \frac{\gamma v^2}{2g}$$

Figures 1.4 through 1.7 illustrate some typical casing entrance and discharge configurations that have evolved through the development of modern centrifugal pumps. However, the impeller inlet and discharge flow conditions will only be ideal when the pump is operating around its BEP, as Figures 1.1 and 1.10 clearly illustrate. In addition to energy efficiency performance considerations, the impeller inlet and discharge flow conditions have pronounced unsteady-flow influences on vibration and noise performance of a pump. This is especially so for high-energy pumps like feed water pumps. Since nuclear power plants are profitable mainly when operated at full rated power output to supply system base-load power, fossil fueled power plants are cycled to provide load following. Under load following, the output flow of major pumps also must be varied over wide flow ranges, particularly at flows below the BEP. Vibrations and pressure pulsations at part load (low flow) pump operation are critical influences contributing to pump component wear, performance degradation, and inoperability/failure sources.

There are essentially two distinct types of flow discharge casings: *volute* and *diffuser*. The two most common volute configurations are illustrated in Figure 1.12. For the single-tongue volute, the static pressure circumferential distribution is nearly uniform only at the design flow, that is, BEP. Therefore

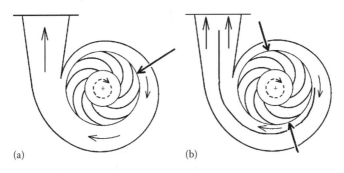

FIGURE 1.12
Volute configuration: (a) single-tongue and (b) two-tongue.

at off-design flows, a circumferentially nonuniform static pressure distribution produces a net static radial force P upon the impeller as illustrated in Figure 1.12a. In the early development period of centrifugal pumps, as speeds and output pressures were being continually increased, it was learned that a significant radial static impeller force was the main reason for high cyclic shaft bending stresses resulting in material fatigue-initiated shaft failures. Stepanoff (1957) was among the first to report on the static radial impeller force in single-tongue volute-type centrifugal pumps Based on numerous centrifugal pump test results in the industry, the magnitude of that static radial force is typically approximated by Equation 1.10 (Adams 2010). The maximum value of K_S depends upon various hydraulic design features, with Stepanoff reporting values for some single-volute pumps as high as 0.6 at shut-off ($Q = 0$) operation.

$$P_s \cong K_s H D_2 B_2 / 2.31 \text{ where, } 144 \text{ in}^2/\text{ft}^2 / 62.4 \text{ lb/ft}^3 = 2.31 \text{ in}^2 \text{ ft/lb} \quad (1.10)$$

where D_2 = impeller OD (inches), B_2 = impeller overall width including shrouds (inches) H = head (feet), P_s = radial resultant impeller force (pounds), and $K_s \cong 0.36 \left[1 - \left(\dfrac{Q}{Q_{BEP}} \right)^2 \right]$ dimensionless coefficient approximation from many test results.

As illustrated in Figure 1.12b, the now common two-tongue volute configuration was devised to have the major portion of the single-tongue impeller force P replaced by two nearly canceling force components. Multiple-tongue volutes, while less common, have been employed by a few manufacturers. In troubleshooting projects, the author has dealt with both equally spaced three-tongue and four-tongue volute configurations.

Volutes simultaneously collect and decelerate the flow exiting from the impeller. The common alternative to the volute approach is the *diffuser* configuration, which first diffuses the impeller discharge flow and then collects it in a nonspiral surrounding chamber. Figure 1.13 shows the essential flow-path comparison between a two-tongue volute and a diffuser configuration. As one would naturally expect, due to the multitude of equally spaced diffuser vanes, it also tends to cancel out the net static radial force upon the impeller. By separating out the in-series diffusion and collection functions, the diffuser approach lends itself to somewhat higher stage efficiency at the BEF. So it has been adopted more commonly in high-energy pumps, for example, feed water pumps. However, at off-design operations, for example, part-load low-flow operation, the diffuser is not quite as energy efficient as the volute approach. That is, at off-design flows the diffuser flow is more "unhappy" than in the less flow constraining, less finicky volute. Furthermore at off-design unfavorable flows, the diffuser is more prone to vane damage accrual than is a tongue in a volute.

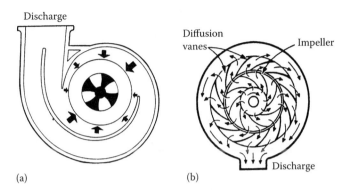

FIGURE 1.13
Comparison of volute and diffuser: (a) two-tongue volute and (b) diffuser.

The fluid within and surrounding the impeller also produces dynamic (time varying) unsteady-flow forces upon the impeller. Available data from laboratory tests for the dynamic impeller forces is presented and delineated by Adams (2010) as comprised of two parts:

- Strictly time-dependent *unsteady flow forces*
- *Interaction forces* produced in response to rotor orbital vibration motion

The interaction dynamic forces are mathematically modeled from test data similar to how journal bearings and dynamic seals are modeled for rotor dynamic analyses, that is, employing a linear dynamics model with stiffness, damping, and virtual mass components as detailed by Adams (2010). The unsteady flow forces are mathematically modeled from test data using an equation similar to that in Equation 1.10 for the static radial impeller force, as follows:

$$K_d(\text{rms}) = \frac{P_d(\text{rms})}{\gamma H D_2 B_2} \text{ Dimensionless with consistent units (SI or English)} \quad (1.11)$$

where D_2 = impeller OD, B_2 = impeller overall width including shrouds, H = head, P_k = radial impeller unsteady flow force, γ = weight density of liquid, and rms is the root-mean-square time averaging.

The severity of unsteady flow impeller forces is strongly influenced by the quality degree of the hydraulic design and manufacturing precision of the impeller. Values for $K_d(\text{rms})$ are given by Adams over a wide frequency range from laboratory test results of a pump stage of very good hydraulic design and a high-precision made impeller, that is, best-case scenario. Figure 1.14 shows an example of $K_d(\text{rms})$ test results as a function of frequency and

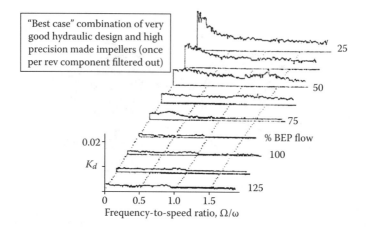

FIGURE 1.14
Spectrum (rms) of broadband impeller unsteady-flow impeller force.

percent of BEP flow. These results clearly explain the large subsynchronous vibration and pressure pulsation amplitudes in high-energy pumps at low flow operation, for example, feed water pumps at part-load power outputs. This obviously explains the standard requirement for feed water pumps to have a part-load bypass flow line to prevent the pump from experiencing flows below typically 25% of Q_{BEP} when the delivered flow is less than 25% of Q_{BEP}. There are many feed water pumps now in service where the manufacturers have compromised the hydraulic design to maximize the peak efficiency a bit for the BEP, but at the expense of producing increased unsteady flow impeller dynamic forces for $Q < Q_{BEP}$. This has required that some bypass flow lines be enlarged, typically to 35% Q_{BEP}.

1.3.2 External and Internal Return Channels of Multistage Pumps

The cross section of a modern *barrel-type boiler feed water pump* illustrated in the top portion of Figure 1.6 is an example of a multistage horizontal pump employing twin-volute internal stage-to-stage return channels. This is an improvement, both structurally and efficiency wise, over the older multistage split-casing configurations employing external return channels like that shown in Figure 1.15. The older configurations with external return channels are typically 3% to 4% less efficient than the newer internal crossovers designs. As an alternative to the twin-volute internal crossover return piece shown in Figure 1.6, some manufacturers employ an interstage piece with diffusers vanes instead of a twin-volute interstage piece like shown in Figure 1.6. Stepanoff (1957) provides design overviews for all these return channel configurations, both for horizontal and vertical centerline pumps.

Current-era installed steam-powered main turbines in the United States are essentially all part of *combined-cycle plants*, for example, the combined

Pump Fluid Mechanics, Concepts, and Examples 21

FIGURE 1.15
Split-casing eight-stage pump with external crossovers.

FIGURE 1.16
(See color insert.) Radially split ring-section feed water pump.

waste heat from three 150 MW combustion gas turbines is used to produce the steam for a 400 MW steam turbine. For these combined-cycle-plant steam turbines, the feed water pumps generally adopted are of the *radially split ring-section* configuration such as shown in Figure 1.16. These pumps are less first-cost expensive than the superior more robust barrel-type feed water pumps.

1.3.3 Pump Priming

Properly sealed positive displacement pumps are self-priming because they work on the principle of pushing a volume and thus can pump out a gas.

Centrifugal pumps on the other hand are not automatically self-priming because they operate on the principle of increasing the pumped medium's angular momentum (Section 1.1.2), thus developing only a small pressure rise when filled with a gas instead of a liquid. For the interested reader, Karassik and Carter (1960) provide an entire chapter devoted to *centrifugal pump priming*. Quoting Karassik and Carter: "A centrifugal pump is primed when the waterways of the pump are filled with liquid to be pumped. Removal of the air, gas or vapor may be done manually or automatically, depending upon the type of equipment and controls used."

1.3.4 Controls

As a trade-off with the automatic self-priming quality of positive displacement pumps, centrifugal pumps are much easier to control than positive displacement pumps. This is because centrifugal pump operating characteristics are quite adaptable to various system head-capacity applications and startup. As in the case of pump priming, Karassik and Carter (1960) also provide an entire chapter devoted to *centrifugal pump controls*. The most important example of this is controlling pump flow. In the case of a centrifugal pump, although the best efficiency is only at one flow point with constant rotational speed, the flow delivered can be simply controlled with a throttle valve, as done with a full array of power plant centrifugal pumps during power *load following* (see Chapter 2, Figure 2.2). On the other hand, positive displacement pumps at constant speed operation require a bypass flow line as illustrated in Figure 1.17. This is because with a fixed volume flow rate of an incompressible fluid, the flow volume has to go somewhere, not being able to simply recirculate within the pump itself. And of course, if it is possible for this bypass value to be closed, there must then also be a pressure relief valve to prevent the serious consequences of starting/running a positive displacement pump with closed shutoff. As with positive displacement pumps, centrifugal pump output can also be controlled by speed control in variable-speed driven pumps.

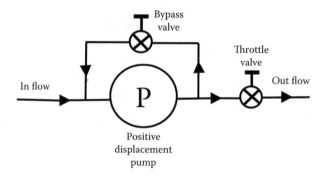

FIGURE 1.17
Elementary control loop for a positive displacement pump.

2

Pump Performance, Terminology, and Components

Power plant pumps are major players both in the overall electric power generation heat-rate efficiency and in plant availability. So the importance of pump power efficiency performance, avoidance of forced outages, and required downtime needed for major overhauls each individually compete for priority with the pump original equipment manufacturer's (OEM) first-cost price tag, at least in the minds of prudent power producers.

2.1 Hydraulic Performance and Efficiencies

In his landmark report, Makay (1978) clearly documented that in high-energy density pumps, too much emphasis had often been placed on maximum pump efficiency at full load, resulting in unfavorable hydraulic performance (e.g., hydraulic instabilities) at part-load operation. Makay reports on numerous troubleshooting examples where severe hydraulic instability (flow surging) caused high levels of pump vibrations and pressure pulsations, leading to seal, bearing, shaft, impeller, and axial thrust-balancer failures. He recommended comprehensive shop-witness tests of new pumps over the entire operating speed and flow ranges, with all performance and reliability parameters carefully measured.

The hydraulic efficiency of a centrifugal pump can be expressed by the ratio of fluid power produced by the pump to the shaft power needed to drive the impeller, as follows:

$$\eta_h = \frac{\text{Fluid power produced by pump}}{\text{Power required to drive impeller}} \times 100\% = \frac{\gamma QH}{T\omega} \times 100\% \qquad (2.1)$$

where T = impeller torque, ω = speed (rad/sec), Q = flow, H = head, and γ = liquid weight density.

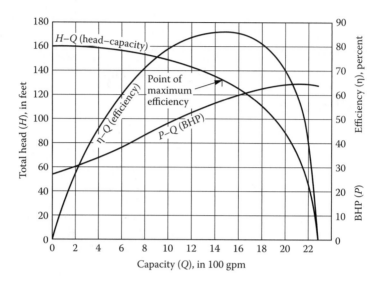

FIGURE 2.1
Pump efficiency, head, and power versus flow at constant speed.

The mechanical efficiency can be expressed as the ratio of shaft power absorbed by the impeller to the total shaft input power, as follows:

$$\eta_m = \frac{\text{Power absorbed by impeller}}{\text{Total shaft input power}} \times 100\% = \frac{T\omega}{P_{\text{shaft}}} \times 100\% \qquad (2.2)$$

where Power absorbed by impeller = Total shaft power − Mechanical losses.

Mechanical losses include those from bearings, seals, and stuffing boxes. Stepanoff (1957) gives additional formulations for these efficiencies, as well as delineations of the sources of hydraulic and mechanical energy losses as functions of flow and specific speed. Figure 2.1 shows a typical plot of pump efficiency η, head H, and power P_{shaft} as functions of flow Q at constant rotational speed.

2.2 Intersection of Pump and System Head–Capacity Curves

A pumped liquid flow loop is analogous to a direct current (DC) electrical circuit, where the pump is like the battery, the flow path resistance is like the circuit electrical resistance, the pump head is like the battery voltage, and the pump flow is like the electrical current produced. So in steady-state operation, the pump's operating point on its head–capacity curve will naturally

Pump Performance, Terminology, and Components

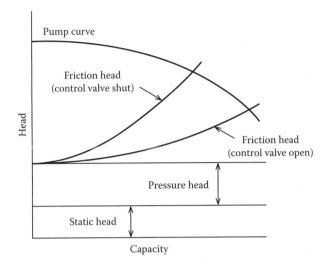

FIGURE 2.2
Intersection of a pump H-Q curve and two different system H-Q curves.

have to match a point on the flow system's head–capacity curve. That is, for steady-state operation the pump H-Q operating point must equal a point on the system H-Q curve. The system head–capacity curve is comprised of two additive parts: the fixed head and the flow-dependent head.

The flow-dependent portion of the system head–capacity curve is generally of a parabolic-like function of flow since it is from the friction resistance to flow in pipes, heat exchangers, and so forth. The fixed head portion can be comprised of two parts: the static head (elevation change under gravity) and an additional imposed static pressure head. Figure 2.2 illustrates this for two system H-Q curves for two different control valve settings. The selection of a pump well suited for a given application starts by matching the pump H-Q curve with the system H-Q curve.

2.3 Cavitation Damage and Pump Inlet Suction-Head Requirements

2.3.1 Description of Pump Cavitation Phenomenon

Cavitation starts with the formation of vapor pockets or cavities (steam bubbles) in any flowing liquid when the liquid flows into a location where the pressure becomes lower than the liquid's vapor pressure. Naturally this fundamental phenomenon is not restricted to pumps. In centrifugal pumps, cavitation is most likely to occur at the inlet (suction) region of the impeller on the low-pressure (trailing) side of the impeller vanes. As these vapor pockets are swept with the flow further into the impeller they experience the progressive increase in liquid pressure that the impeller naturally produces. These vapor pockets will therefore collapse inside the impeller. But because the transient from vapor back into liquid involves thermodynamic heat transfer, the collapse of the vapor pockets does not occur immediately upon experiencing a pressure that just exceeds the vapor pressure. The very small but finite thermodynamic time delay for the vapor pockets to collapse means that the collapse will occur at points where the local pressure has already exceeded the vapor pressure by a significant amount. In consequence the collapse of the vapor pockets creates violent implosions that act to progressively erode the impeller vane surface if the pump is operated with insufficient suction pressure to disallow the initial formation of vapor pockets. Collapse of the bubbles is nonspherical, being more correctly likened to an intense microjet. Karimi and Avellan (1986) provide an in-depth insightful presentation of their fundamental research on cavitation.

When the intense microjet is closely directed into the vane surface, it contributes significantly to vane surface erosion. Figure 2.3 is a photo of a centrifugal pump impeller vane inlet that has incurred substantial cavitation erosion damage. Naturally, such damage adversely affects pump energy efficiency as well as impeller structural integrity.

Cavitation caused erosion can also occur on hydro turbine vanes at the discharge low-pressure region of the turbine impeller. This can occur for example when a turbine is operated at output power levels exceeding the

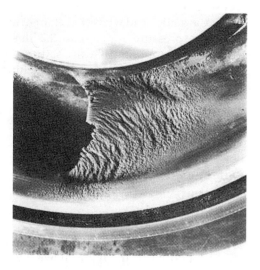

FIGURE 2.3
Cavitation erosion damage of an impeller vane inlet.

design power rating of the turbine, feasible if the unit's generator is rated above the turbine's power rating. Such an operating point above the turbine's rating might be chosen during the seasonal spring heavy run-off river flow. The author has witnessed examples of this at hydroelectric plants employing axial-flow Kaplan turbines where the generating income from the extra power so produced substantially exceeded the cost of periodically stainless-steel recladding the cavitation-damaged portion of the turbine impeller vane surfaces.

2.3.2 Laboratory Shop Testing to Quantify Pump Cavitation Incipience

Causes of pump cavitation primarily include inadequate *net positive suction head (NPSH)*, that is, not enough pressure at the pump inlet; flow recirculation at the impeller eye while operating at off-design flows (see Chapter 1, Figure 1.1); incorrect vane inlet angle β_1; and localized high fluid velocities caused by sharp corners and/or misplaced and blunt inlet guide vanes (Makay 1978).

Required NPSH is a basic characteristic of a pump stage and thus is determined as part of the controlled testing of any new hydraulic design. The standard test to quantify the required NPSH for a pump stage entails holding constant the head produced by the stage for a given flow and slowly reducing the suction head until the head produced exhibits a sharp decline; that is, slowly reducing both the discharge and inlet heads simultaneously at the same incremental amount, maintaining their difference constant. This slow reduction in NPSH is continued until the pump exhibits a pronounced decrease in performance by a drop in head and/or drop in the experimentally determined head-ratio coefficient σ of Thoma (1937), which is based on flow dynamic similitude of geometrically homologous stages operating at the same specific speed. Thoma's coefficient is defined as follows (see velocity inlet triangle in Chapter 1, Figure 1.8):

$$\frac{c_1^2}{2g} + \lambda \frac{w_1^2}{2g} = \Delta h = \sigma H \tag{2.3}$$

Figure 2.4 shows a graph of test points from a typical controlled variable-NPSH test to determine the minimum NPSH required to avoid cavitation for the tested H-Q operating point.

Traditionally, the required minimum NPSH has been defined by the pump industry to be where the performance drop is 2% of the performance illustrated in Figure 2.4. However, Makay highly recommended in many of his publications that, especially on high-energy pumps like for feed water, 2% is too high and is thus insufficiently conservative, preferring instead the lowest value perceptible from careful interpretation of the test results.

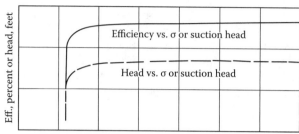

FIGURE 2.4
Pump performance to find required NPSH at constant speed and flow.

2.3.3 Required Net Positive Suction Head and Available Net Positive Suction Head

As already stated in Section 2.2, the selection of a pump well suited for a given application should start by matching the pump H-Q curve with the system H-Q curve, as shown in Figure 2.2. In addition and of equal importance is making sure that the available NPSH at the application site comfortably exceeds the minimum required NPSH of the pump. Figure 2.5 shows a typical graph of the required NPSH as a function of flow combined with the pump head–capacity and efficiency curves for two different types of suction nozzles. The larger the margin between the available and required NPSH, the greater is the margin of protection from cavitation that can occur even when the available NPSH exceeds the required NPSH,

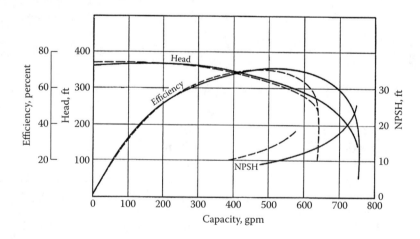

FIGURE 2.5
Example of required NPSH at constant speed for two different suction nozzles. Straight suction nozzle (straight line), flat elbow suction nozzle (dashed line).

for example, from inlet flow distortions at the impeller eye as itemized in Section 2.3.2.

The presence of pump cavitation can be detected by measuring the resulting fluid-borne noise with piezoelectric pressure transducers or from solid-borne noise measured with accelerometers. The strength of these cavitation noise signals can be especially strong in high-energy pumps, depending on the severity of the ongoing cavitation. When significant pump cavitation is detected from noise measurements and/or erosion damage of impeller vane inlets (see Figure 2.3), injection of air into the pump suction zone can mitigate the damage potential of the cavitation. Such continuous air injection during operation acts as a "cushion" to lessen the severity of the violent vapor pocket collapses described in Section 2.3.1.

2.3.4 Operation of Pumps in Parallel

There are a number of reasons why parallel operation of two or more pumps is a preferred alternative to using one sufficiently large pump alone to produce 100% of an application's flow requirements. In a common circumstance illustrated in Figure 2.6, two pumps in parallel can allow continued system operation, albeit at a reduced capacity, when one of the two pumps needs to be taken out of service, for example, due to symptoms of impending failure or unexpected outright failure to operate. That is why most power plant feed water systems have two 50% pumps. Because of the parabolic-like shape of the pipe-like friction portion of the feed water system H-Q curve, operation out to approximately 65% of the feed water system's rated capacity is feasible with just one 50% feed water pump (see Figure 2.2). So the main steam turbine-generator unit can produce approximately 65% of its rated capacity with only one 50% feed water pump operating. It is important that pumps operated in parallel have near identical H-Q curves, otherwise the total flow will not be equally divided between the pumps.

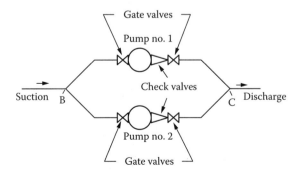

FIGURE 2.6
Piping schematic for two pumps operating in parallel.

2.4 Mechanical Components

2.4.1 Shafts

As with rotating-machine shafts in general, a pump shaft carries all the rotating parts that are attached to it. Pump shafts therefore carry at least the impeller(s), even for the simplest least demanding centrifugal pump applications (e.g., swimming pool). Power plant pumps are among the most demanding of pump applications. And the most demanding among power plant pumps are perhaps the feed water pumps. Those pumps have shafts that carry not only the impeller(s), but various shaft sleeves, rotating parts of dynamic seals and thrust-balancers, and coupling components. The confluence of static and dynamic forces (see Figure 2.7) imposed upon a high-energy pump shaft demand concerted engineering attention to its design.

Even with the best of design practices, the torturous duties of high-energy pumps are so pervasive that shaft failures still occur. Shaft failure of course causes a forced outage of the pump as well as reduced power output of the generating unit. Figure 2.8 shows the most frequent locations for shaft failures in multistage high-energy pumps.

Vibration characteristics are a strong function of shaft geometry. As described in "Troubleshooting Case Studies," Section III of this book, pump rotor vibration is both a revealing symptom for a number of different pump operating problems as well as a concern itself when vibration levels exceed recommended maximum allowable values.

Clearly, proper design of the shaft, for example, correct shrink fits, no square shoulders, manufacturing procedures, heat treatment, conscience quality assurance, power torqueing of retaining nuts, and correct alignment, are all very important in preventing shaft failures.

2.4.2 Couplings

The two basic coupling types used to connect pumps and drivers are the *rigid adjustable* and the *flexible*. Dufour and Nelson (1993) provide a quite complete treatment of all the coupling configurations in use for connecting pumps to their drivers. Close-coupled vertical-centerline pumps utilize a rigid adjustable coupling that transmits axial load up or down to the motor shaft. The coupling thus combines the motor and pump shafts into a single shaft as illustrated in Figure 2.9. Since the weight and axial hydraulic forces of a vertical rotor pump are carried in the axial thrust direction by the thrust bearing(s), the radial bearing support points do not compete for determining pump-to-driver alignment to the large extent encountered by horizontal rotors. Thus horizontal rotors usually require a *flexible coupling* to accommodate the unavoidable radial misalignments that accrue from assembly tolerances, differential thermal expansions, support structure shifting, and so on.

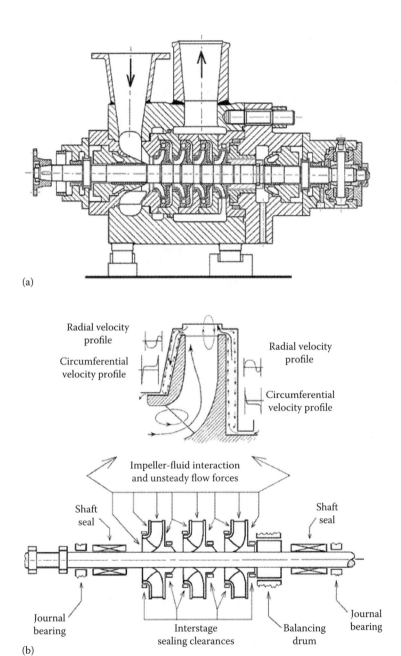

FIGURE 2.7
Multistage boiler feed water pump. (a) Pump cross section and (b) sources of interaction and unsteady flow rotor forces.

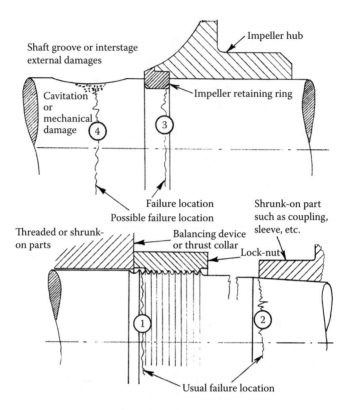

FIGURE 2.8
Multistage pump frequent shaft failures, order numbered by frequency.

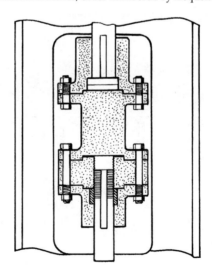

FIGURE 2.9
Rigid adjustable coupling for vertical pumps.

Pump Performance, Terminology, and Components

There are a number of mechanical flexible coupling types of which *gear couplings* are the most common, typified by the cutaway photo in Figure 2.10. Gear couplings have high torque capacity and are relatively compact. But due to the oscillatory relative slipping between the mating teeth, lubrication by oil or grease is required. This is the major disadvantage of gear couplings since inattention to this by maintenance personnel can result in sudden seizure, causing shaft breakage. The photo in Figure 2.11 shows the destroyed

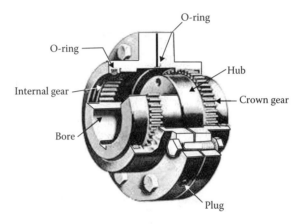

FIGURE 2.10
Gear coupling.

FIGURE 2.11
(See color insert.) Pump shaft after sudden seizure of lubrication-starved gear coupling.

shaft of a 14,500 hp boiler feed water pump caused by the sudden seizure of the gear coupling due to the absence of adequate lubrication.

The *grid coupling* is a competitor to the gear coupling. It has roughly half the torque capacity per unit of weight as the gear coupling. But it shares the disadvantage of the gear coupling in that it also needs lubrication. Figure 2.12 shows two common configurations for the grid coupling.

Flexing-element couplings utilize an elastomeric flexure element or a metallic flexure element; an example of each is illustrated in Figure 2.13. Both renditions eliminate the need for lubrication, a significant advantage over the gear and grid coupling types. The elastomeric flexure design utilizes low-strength materials, which means the size and weight increase considerably with increase in torque requirements. The metallic flexure element design accommodates both angular and translational centerline-to-centerline radial-offset misalignments, but can accommodate only about half of the gear coupling offset. Flexing-element couplings are typically heavier than a gear coupling of the same torque rating.

FIGURE 2.12
Two typical grid coupling configurations.

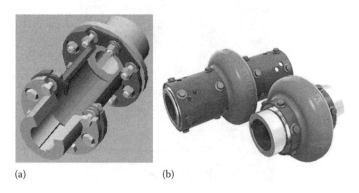

FIGURE 2.13
Flexing element coupling. (a) Metallic flexure element and (b) elastomeric flexure element.

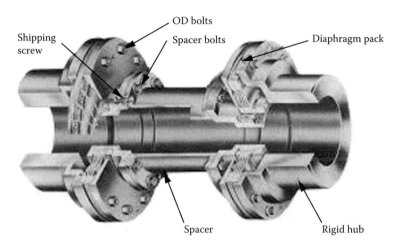

FIGURE 2.14
Diaphragm couplings.

For high-energy units like feed water pumps, Makay (1978) strongly recommended the *diaphragm type coupling*, with the flexibility provided by an axial pack of one or more diaphragms (Figure 2.14). Flexing of the diaphragm element(s) provides both angular and translational centerline-to-centerline offset misalignments. They have high torque capability and do not require lubrication.

2.4.3 Bearings

Bearings position the rotor and thus are the primary load-support points for the rotor. They also are major players in the *rotor vibration* behavior of a pump. There are basically two general types of bearings employed for pumps: *rolling-contact* (usually ball bearing), often erroneously called anti-friction bearings; and *fluid-film* (usually hydrodynamic). Utilizing a rolling-contact bearing always entails choosing them from a bearing manufacturer catalog because of the specialized high precision manufacturing with very fine contact surface finishes. In contrast, fluid-film bearings are often designed and produced by the pump manufacturer, although there are manufacturing suppliers of fluid-film bearings.

Rolling-contact bearings are all finite-life rated as rigorously instructed in the manufacturers' catalogs and well covered in a devoted chapter of any standard undergraduate mechanical engineering machine design textbook. The mechanism that limits rolling-contact bearing life is *subsurface initiated high-cycle material fatigue*, which is caused by the periodic intermittency of the ball or roller contact surface loads traveling over raceway surface points of the bearing (see Figure 2.15). The associated fatigue crack starts in a raceway subsurface plane of maximum alternating shear stress per the classical

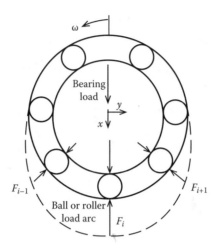

FIGURE 2.15
Typical distribution of contact loads in a rolling-contact bearing.

Hertzian contact stress theory. This failure process culminates with small but macroscopic surface flakes of raceway material detaching (called spalling). However, even though subsurface initiated fatigue is the ultimate life limiter of rolling-contact bearings, in many instances the bearing life is cut shorter for other reasons such as improper installment setting, inadequate lubrication, hard particle dirt ingestion, moisture, and corrosion.

Hydrodynamic journal and axial thrust bearings are designed to operate under load with a small but finite minimum lubricant film thickness without rubbing contact. Theoretically, solving of the Reynolds lubrication equation (Adams 2010) suggests "infinite life." But, of course, some combination of starts and stops, over loads, hard particle dirt ingestion, starved lubrication, excessive journal-to-bearing axial misalignment, and so forth will keep a fluid-film bearing from lasting "forever." Yet there is no well-defined life limiter as there is for rolling-contact bearings. Figure 2.16 illustrates the basic 360° cylindrical journal bearing with pressure distribution and nomenclature. W is the static load upon the bearing; F is the instantaneous fluid film force upon the journal, which includes the static portion $-W$ plus the dynamic reaction to journal orbital vibration.

Pivoted (or tilting) pad journal bearings (PPJB) have been frequently utilized by pump designers and as retrofit improvements in high-energy high-speed pumps, because of PPJBs' superior rotor vibration characteristics. But they require informed technical understanding by the pump designer to avoid their misapplication. With the illustrations in Figures 2.17 and 2.18, Adams (2010) explains the function and proper application of PPJBs.

Ball bearings readily support both radial and axial thrust loads. Even *radial-contact* ball bearings can support some modest thrust load. For simultaneous support of substantial radial and thrust loads, angular-contact ball

Pump Performance, Terminology, and Components

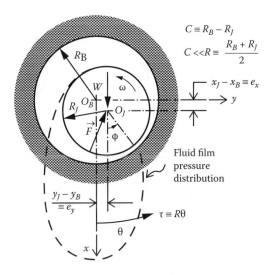

FIGURE 2.16
Hydrodynamic journal bearing and nomenclature; clearance exaggerated.

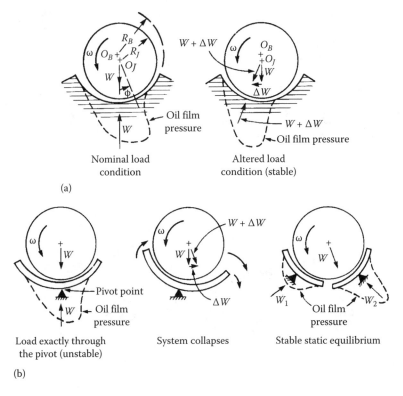

FIGURE 2.17
Comparison between (a) cylindrical and (b) PPJBs.

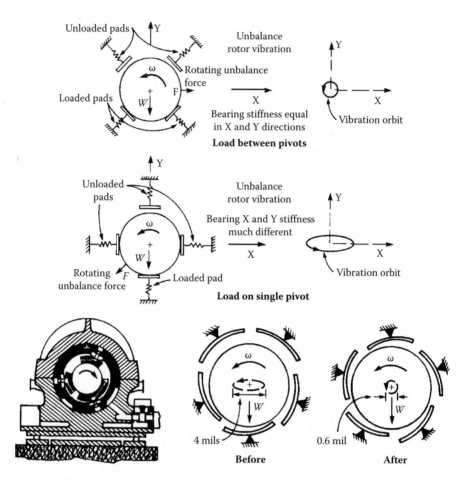

FIGURE 2.18
Load-direction vibration factors of PPJBs.

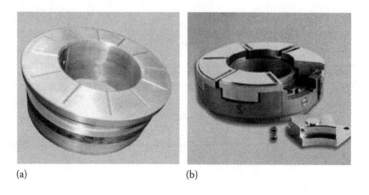

FIGURE 2.19
Hydrodynamic thrust bearings. (a) Fixed profile and (b) self-aligning tilting pad.

bearing configurations are well suited. However, when employing hydrodynamic fluid-film bearings, a separate axial thrust bearing is required, as is obvious from Figure 2.16, which shows that a journal bearing's load capacity is limited to the radial direction. That is, the journal is free to translate axially with respect to the journal bearing. Two frequently employed hydrodynamic axial-thrust bearing configurations are shown in Figure 2.19.

2.4.4 Seals

Like turbomachinery in general, centrifugal pumps need shaft seals to control leakage from inside the pump to the outside, as well as to control interstage leakage between stages within multistage pumps. Figure 2.7 locates both of these pump-sealing functions. Buchter (1979) is a most comprehensive reference on industrial sealing. The seals employed for centrifugal pump needs are from a multitude of types and configurations. Figure 2.20 categorizes the total domain of seal types starting with the two major categories: *static seals* and *dynamic seals*. Static seals refer to sealing between two parts that are not in relative sliding motion, whereas dynamic seals refer to sealing between two parts that are in relative sliding motion. The specific seal category in Figure 2.20 that pertains to centrifugal pumps is *seals for rotating shafts*. Figure 2.21 further delineates *rotating shaft seals*. However, some of the dynamic seal types for rotating shafts incorporate static seals, such as O-rings. To zero in on seals used specifically for centrifugal pumps, Dufour and Nelson (1993) is quite complete, especially for pump users' needs.

Smooth bore cylindrical bushings are frequently used to control leakage at the eye of the impeller and are usually referred to as *wear rings*. That is, they minimize the portion of impeller discharge flow that escapes back into the impeller inlet flow, a direct efficiency loss. Smooth bore bushings are also frequently used to control leakage at the high-pressure side of the impeller to minimize leakage from the next downstream impeller inlet in a multistage pump. Pressure at the next downstream impeller inlet is higher than the

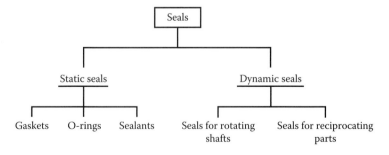

FIGURE 2.20
General seal categories.

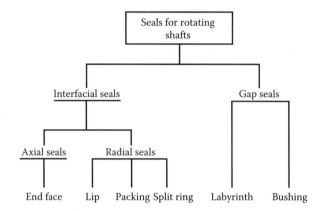

FIGURE 2.21
Rotating shaft seals used in centrifugal pumps.

pressure at the upstream impeller discharge, because the upstream impeller discharge flow is diffused before reaching the downstream impeller (see Figure 2.7). However, the pressure drop across an interstage bushing is considerably smaller than that across the impeller eye wear ring. The diffusion/collection region of the pump (volute or diffuser) is discussed in Chapter 1, Section 1.3.1 (i.e., $\Delta p = \gamma v^2/2g$). Three commonly used versions of the smooth bore bushing annular seal are illustrated in Figure 2.22.

In addition to controlling impeller-eye and interstage leakages, smooth bore bushings also produce a rotor centering force that is (a) proportional to the pressure drop across the bushing, (b) proportional to the rotor-to-bushing radial eccentricity, and (c) inversely proportional to the radial clearance. This is called the *Lomakin effect*, explained in detail by Adams (2010). Over a period of time the bushing clearances naturally enlarge due to rubbing wear, which not only degrades pump efficiency but also proportionally reduces

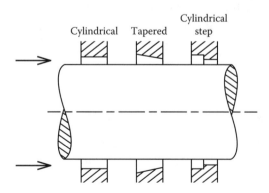

FIGURE 2.22
Smooth bore annular seals. The arrows represent the direction of flow (clearances exaggerated).

the beneficial rotor-dynamic centering stiffness Lomakin effect. It is widely recommended that when these bushings wear open to about twice their as-new value, they should be replaced. Thus the three reasons for this recommendation are as follows:

1. Return the pump to its as-new efficiency.
2. Reduce the amount of impeller inlet flow distortion caused by the impeller eye wear ring leakage mixing back into the primary impeller inlet flow.
3. Prevent a rotor vibration critical speed originally above the maximum operating speed from dropping into the operating speed range.

In some modern designs of high-energy pumps, specific efforts to improve pump vibration behavior by optimizing bushing geometry to increase their radial stiffening capability has been foiled when the bushings wear open and a critical speed drops into the operating speed range. The as-new radial clearances are typically twice the journal bearing radial clearances to ensure the bushing seals do not inadvertently act as primary bearings. However, in some vertical multistage pumps with quite long skinny shafts, the interstage bushings do in fact become bearings.

Grooved configurations (Figure 2.23) are an alternative for the same impeller controlled leakage functions just described for smooth bore bushings. Deep grooved configurations are more forgiving to rotor-stator rubs, that is, less likely to seize under extreme rub conditions. However, while their

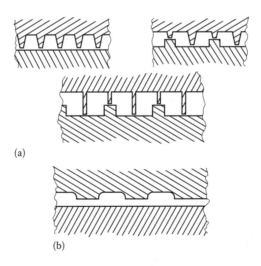

FIGURE 2.23
Examples of circumferentially grooved annular seals. (a) Labyrinth seals; groove depth much larger than radial tip clearance. (b) Shallow-grooves; groove depth approximately equal to tip clearance.

leakage control capability is comparable to smooth bore bushings, they have virtually no Lomakin effect, that is, no radial stiffening effect. Makay pioneered the use of the shallow groove configuration (Figure 2.23b), which retains some of the Lomakin effect (flat tooth tips) and retains some of the rub forgiveness of the sharp-tooth-tip labyrinth (Figure 2.23a).

Preventing the pumped liquid from leaking inside the pump to outside the pump is more than just a pump efficiency performance matter, but more important it is often a containment and safety matter. *Packing* is the simplest and oldest approach to pump shaft sealing. Figure 2.24 shows a typical packing-type shaft seal. The leakage for this seal type is not zero but is quite small, adjusted by the packing gland to keep the shaft-rubbing packing rings lubricated by the pumped liquid. The *packing gland* is an axially positioned-adjustable component used to provide the correct compression force on the packing rings to seal properly while still keeping them adequately lubricated.

Mechanical face seals are well suited to many power plant applications like feed water and other high-pressure pumps. There are many different configurations for *face seals*. A basic configuration is illustrated in Figure 2.25. The heart of the face seal is where the seal stator and seal rotor meet face-to-face. That is where the seal pressure drop takes place. The spring is employed to keep the faces in contact with a sufficient force to prevent the faces from coming out of mating in operation. In order for this mechanism to work, the mating surfaces must be extremely flat, typically within one helium light band. The author has often heard, with some amusement, face-seal "experts" state why these face seals work so well with such extremely thin film thicknesses. They say it is because in manufacturing the mating surfaces to be as close as possible to perfectly flat, they never of course quite

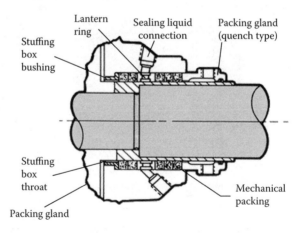

FIGURE 2.24
Typical packing shaft seal.

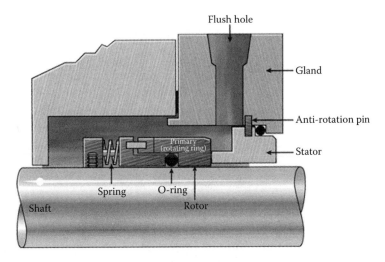

FIGURE 2.25
Basic mechanical face seal.

get them perfectly flat. To the author's knowledge, the most comprehensive treatment of face seal technology is that of Lebeck (1991).

A fluid-film variation of the face seal concept is employed in the pressurized water reactor (PWR) primary coolant pump in Figure 2.26. It is the *tapered-land* fluid film seal that incorporates a radially convergent film thickness in the flow path direction. It was developed in the early years of commercial nuclear power by a major manufacturer of PWR primary-loop coolant pumps. The primary-loop radioactive water nominal pressure is about 2250 psi (153 atm). The tapered-land seal operates with a "nonzero" film thickness less than 0.001 inch (40 micrometers), that is, a true fluid film seal. With this finite film thickness and very hard aluminum oxide face inserts, it was developed to provide the maximum insurance against seal failure in this safety related nuclear pump. It is, however, backed up by a secondary conventional face seal that is designed to take the full shaft sealing mission in case the primary tapered-land seal fails. In normal operation, the leakage flow through this shaft seal is drained to the inlet of the reactor charging pumps employed to control rector primary-loop pressure. The tapered portion of the seal stator face acts to "bulge out" the pressure-drop distribution across the seal face as the film thickness is reduced (a hydrostatic stiffening property). A later derivative of this approach is the *stepped-land* seal, which has a similar hydrostatic stiffening property. Makay et al. (1972) provide detailed design information and procedures for the tapered-land seal.

Floating-ring shaft seals are often employed for high-pressure pumps in the petrochemical industry. Childs and Vance (1997) present a comprehensive technical treatment on floating-ring seals. This concept also made

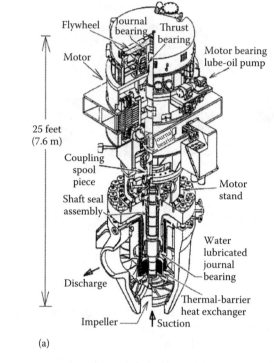

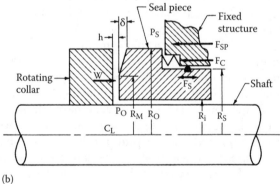

FIGURE 2.26
PWR primary coolant pump and its tapered-land primary shaft seal. (a) PWR primary coolant pump and (b) tapered-land fluid film seal.

its way into some high-pressure power plant pumps, in particular boiler feed water and boiler circulating pumps. There are several shaft seal configurations marketed that utilize the floating-ring concept. As treated in Sections II and III of this book, floating-ring seals have shown themselves to be wear-prone and fragile in the high-pressure water pumps of power plants. Figure 2.27 shows a typical configuration employing multiple floating rings.

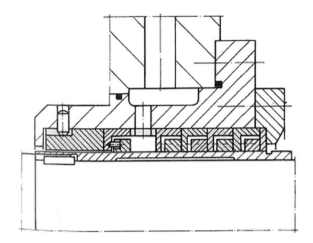

FIGURE 2.27
Floating-ring shaft seal.

2.4.5 Thrust Balancers

Dufour and Nelson (1993) provide comprehensive treatment of centrifugal pump rotor axial loads, from small single-stage pumps to large multistage high-energy pumps. High-pressure multistage pumps typically can have very high hydraulic axial forces that load the rotor beyond the load capacity of any conventional thrust bearing. The source of axial hydraulic rotor force is illustrated in Figure 2.28 for a single stage with optional features to achieve axial hydraulic force balance. The total axial thrust load upon a multistage rotor is the summation of the thrust contributions from all of the stages. In the early development of high-energy multistage pumps, the approach to handle hydraulic thrust was axially opposed stages to balance out the

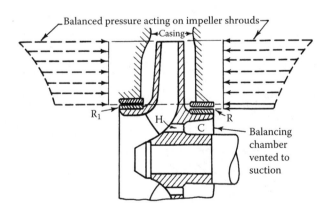

FIGURE 2.28
Single-stage pump with back wear ring and holes for axial balance.

nominal rotor thrust (Figure 2.29). This approach necessitates an external crossover pipe to connect the two intermediate stages, which comes with an efficiency penalty and a manufacturing incremental cost increase. Therefore, large high-energy multistage pump design evolved to balance out the net rotor thrust load by other means.

For high-energy pumps, the two main approaches now used are the *balance drum* and the *balance disk*. The pump illustrated in Figure 2.7 has a balance drum. But it still has a double acting thrust bearing, shown on the right side of the illustration. The same would be the case for the other two options: opposed stages or balance disk. Since none of these three approaches will perfectly balance the net thrust load, a thrust bearing is needed to carry the relatively small residual thrust and to axially position the rotor. The same is true for large multirotor steam turbines that employ both double-flow rotors and balancing drums, but still need a double-acting thrust bearing to axially position the rotor. For pumps or turbines, the rotating impeller vanes or turbine blades must axially line up accurately with their respective nonrotating partners. It should come as no surprise that when a balance disk or balance drum fails, the thrust bearing is quickly destroyed, as are probably all or most of the pump internals.

The basic geometry of balance drums and disks are illustrated in Figure 2.30, with high pressure at the left sides on the illustrations and low pressure at the right sides. These balance devices are subjected to approximately the pump full discharge pressure on one side and suction pressure on the other side. With these pressures, diametrical dimensions are selected to obtain

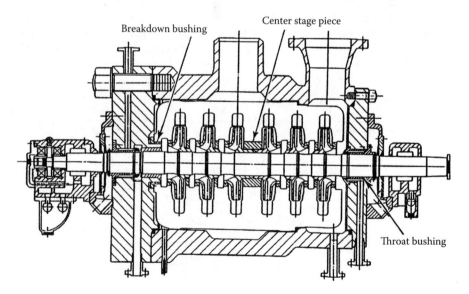

FIGURE 2.29
Multistage pump with axially opposed stages.

Pump Performance, Terminology, and Components

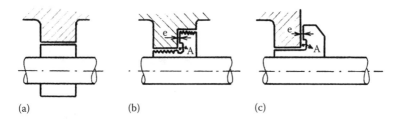

FIGURE 2.30
(a) Basic balance drum and (b, c) disk configurations.

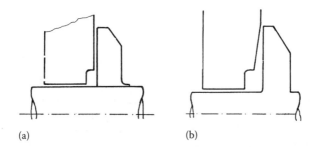

FIGURE 2.31
(a) Parallel face and (b) Makay-taper face balance disk configurations.

the needed annular surface areas to balance the combined axial thrust from all the stages. Two options for the balance disk approach are illustrated in Figure 2.31. The Makay-taper option is the superior design because its separating force increases all the way to closure, in contrast to the parallel-face option. That is, it is conceptually similar to the PWR primary coolant pump tapered-land primary shaft seal illustrated in Figure 2.26.

2.5 Drivers

The pump drive can be a constant speed electric motor coupled directly or through a gear box, in which case the delivered flow is controlled by a discharge valve on the pump discharge line as shown in Figure 2.2. The pump then operates at its best efficiency point (BEP) only at its maximum efficiency design-point flow as the example constant-speed efficiency curve in Figure 2.1 demonstrates. Another constant-speed option is to have a 100% boiler feed pump direct connected to the main steam turbine shaft. With this option, electric motor driven boiler feed start-up pumps are needed to start the boiler and thus to get the main steam turbine started. The pump can also

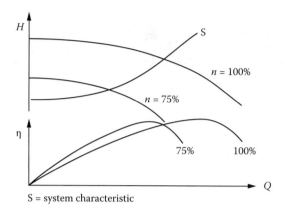

FIGURE 2.32
Efficiency (η) and head–capacity (H-Q) with speed (n) variation.

be motor driven through a variable speed hydraulic coupling or auxiliary steam turbine. The decision on whether to employ an electric motor or turbine drive is based on the power, with electric motors generally the preference up to about 14,000 hp (Makay 1978).

Variable speed drives allow a pump to operate near a BEP over its full operating flow range, controlling pump flow by controlling the pump rotational speed. This is demonstrated by the H-Q and η curves at two different rotational speeds as shown in Figure 2.32. The largest capacity boiler feed pumps in service in the United States have two 50% pumps, each turbine driven and rated at 80,000 hp, supplying feed water to the 1300 MW Brown-Boveri cross-compound steam turbine generating units. Motor-driven boiler feed start-up pumps are also required for this variable speed option.

As just stated, a variable speed option allows operation close to a BEP over the full operating flow range by setting flow through controlling speed. This advantage also lessens the impeller inlet velocity incidence angle, thus lessening flow separation and consequently lessening unsteady flow impeller dynamic excitation forces. This advantage of variable speed drives consequently lowers dynamic stresses in various pump components, and lowers the risk of cavitation erosion because of lower impeller tip speed. Consequently, variable speed drives increase the life of pump components and correspondingly reduces maintenance demands.

More often, large generating units, both fossil and all nuclear, utilize two 50% feed water pumps so that the unit can still generate at approximately 65% of its full-rated power output when one 50% feed water pump is out of service. An example of a large power plant pump that is electric motor driven at constant speed is the nuclear PWR primary coolant pump like the one shown in Figure 2.26 (100,000 gpm at 1200 rpm).

3

Operating Failure Contributors

Coping with power plant pump operating problems is a way of life for plant operators. Avoidance of pump forced outages, downtime, and time between major overhauls are all top priorities for maximizing power plant availability. Makay (1978), referenced also throughout Chapter 2, focused on feed water pumps because of their demanding high-energy-density service and resulting vulnerability to numerous failure mechanisms. However, other major pumping functions are also critical to plant availability, being vulnerable to many of the same as well as other failure causes.

3.1 Hydraulic Instability and Pressure Pulsations

Unstable pump flow and troublesome pressure pulsations while related to each other are distinct phenomena. A full technology assessment and an understanding of these phenomena are embedded in quite complex 3D unsteady-flow fluid mechanics. The objective of the presentation here is primarily to provide plant engineers an appreciation for these phenomena, not to supply a research thesis on unsteady flow.

3.1.1 Head–Capacity Curve Instability

The shape of a centrifugal pump's head–capacity H-Q curve has a major impact on its operability. The best H-Q curve is one where the head continuously rises as the flow is progressively reduced to zero as illustrated in Figure 3.1a. A *drooping H-Q curve* is shown in Figure 3.1b. Truth be known, the H-Q test results of many centrifugal pumps droop a bit at low flows, even though pump manufacturers' published H-Q curves generally do not show that. But as long as a centrifugal pump is not operated too near the zero-flow condition, such "cosmetic doctoring" of the manufacturer's H-Q curve near shutoff will not lead to operating problems.

A pump with a drooping H-Q curve as shown in Figure 3.1b may exhibit unstable flow (surging) at flows below the peak of the H-Q curve. The additional condition needed to cause this self-excited unsteady flow is a compressible volume somewhere in the system, for example, water in long pipes, or steam volume in the deaerator or boiler. This is similar to the characteristic

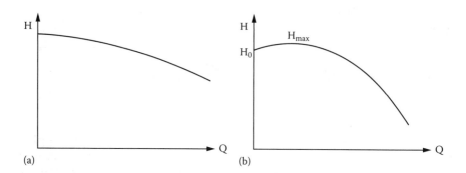

FIGURE 3.1
Head–capacity curves: (a) rising and (b) drooping.

of all centrifugal compressors that encounter a surge line as flow is reduced, because the fluid (i.e., gas) itself is a compressible volume. While the full nature of such pump surging unsteady flow is quite complicated, a heuristic way of viewing it is that there are two Q values having one H value, so the pump "can't make up its mind" which of the two Q values to pick with the H fixed by the system H-Q curve. Graphing the drooping (positive slope) portion of a pump H-Q curve as a continuation of the more steady flow rising (negative slope) portion of the H-Q curve is misleading because the behavior of pump flow to the left of the pump H-Q peak may be highly unsteady.

In addition to the possibility of unstable pump flow phenomena occurring with a drooping H-Q curve, it is also possible with an S-shaped (saddle) H-Q curve (Figure 3.2). In this case, the possibility of *stable or unstable pump flow* depends upon the relative slopes of the intersecting system and pump H-Q curves. Figure 3.3 illustrates these two possibilities. Figure 3.3a shows the stable case, with the system H-Q curve having a higher slope than the pump H-Q curve. Figure 3.3a demonstrates that with a momentary flow perturbation ΔQ, the corresponding perturbation in the system head ΔH_{sy} is larger than the corresponding perturbation in the pump head ΔH_{pu}. That forces

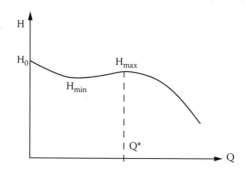

FIGURE 3.2
Example of a pump saddle-shaped H-Q curve.

Operating Failure Contributors

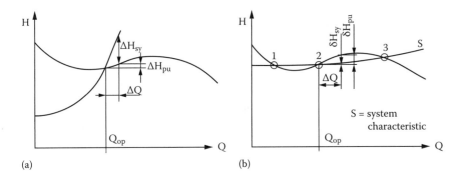

FIGURE 3.3
Pump H-Q intersection with system H-Q: (a) stable and (b) unstable.

the flow point back to the intersection of the two curves, and likewise for a negative flow perturbation.

In contrast, Figure 3.3b shows three intersection points of the pump and system H-Q curves, where the intersection point 2 indicates an unsteady flow condition since any flow perturbation ΔH at 2 will cause the flow at that point to shift to one of the two stable flow points, 1 or 3, where the system curve has a higher slope than the pump curve. Figure 1.14 (see Chapter 1) shows a graph from experimental research results of impeller dynamic unsteady hydraulic forces upon an impeller as a function of percent of best efficiency flow and as a function of the ratio of frequency to rotational-speed frequency (Adams 2010). As Figure 1.14 shows, the unsteady fluid dynamic impeller forces get stronger the lower the flow from the best efficiency point (BEP), with the dominating dynamic forces at frequencies predominantly below the rotational frequency. This is consistent with many pump vibration measurements taken on vibration-troubled power plant pumps.

3.1.2 Pressure Pulsation Origins

Flow recirculation occurs both at the impeller inlet and discharge. Centrifugal pump technologists are aware that even at the best efficiency flow, a modest amount of flow recirculation is a good thing because it is the flow's way of adjusting to inherent imperfections in the pump geometry's difficult job of efficiently adding mechanical energy to the liquid. But when a pump is operated at flows considerably away from its BEP, the strong unsteady recirculation that naturally ensues brings with it considerable flow unsteadiness with large pressure pulsations, noise, and excitation forces (see Chapter 1, Figures 1.1 and 1.14). Large-scale flow vortices and fluctuating lift are responsible for this.

Wake flow at the impeller discharge is a significant contributor to pressure pulsations. Specifically, (a) finite vane thickness at the trailing edge, (b) boundary layers on both sides of the vane, and (c) velocity distribution

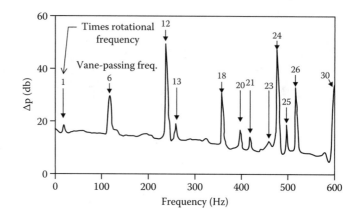

FIGURE 3.4
Discrete-frequency spectrum of pressure pulsations at pump discharge. Number of impeller vanes = 6, number of diffuser vanes = 11.

differences between the pressure and suction sides of the vane, all create wakes that feed pressure pulsations.

Several documented successful power plant troubleshooting case studies have demonstrated that (a) the radial gap between impeller shrouds and diffuser vane tips (Gap A), and (b) the radial gap between impeller vane tips and diffuser vane tips (Gap B) (see Figure 3.18) strongly influence

- Stability of the *H-Q* curves
- Steady and unsteady axial thrust on the impeller
- Pressure pulsations

This is well documented, for example, by Makay (1980), Makay and Barrett (1984), and Guelich et al. (1989).

Broadband and discrete-frequency pressure pulsations are a function of the particular hydraulic design, rendering their accurate prediction not possible. Thus experimental results are the only option for a particular pump operating in its particular system (Guelich and Bolleter 1992). The nature of broadband pressure pulsations is reflected in the broadband impeller force measurements presented in Figure 1.14. An example of measured discrete-frequency pressure pulsations is shown in Figure 3.4.

3.1.3 Criteria for Minimum Recirculation Flow

The operating problems and associated pump failures resulting from hydraulic instability and pressure pulsations are most activated when a pump is operated at flows *significantly below the best efficiency flow*. As well known, fossil-fired plants are commonly used now for load following

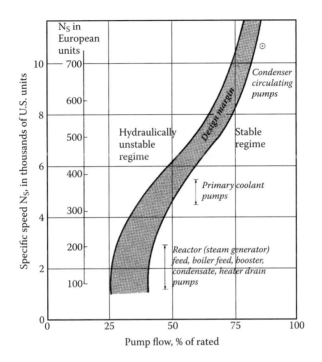

FIGURE 3.5
Makay minimum bypass flow guideline for centrifugal pumps.

because the nuclear plants are best suited to be run at their full-rated capacity for base load power demand. This has accentuated the attention given to the required minimum-allowable bypass (recirculation) part-load pump flow. Through his pioneering troubleshooting of power plant pumps, Makay (1994) established guidelines covering the full spectrum of power plant pump applications for minimum bypass flow requirements, which are shown in Figure 3.5.

3.2 Excessive Vibration

This section is written for anyone with an elementary grounding in vibration fundamentals, in particular the classical one-degree-of-freedom model. In addition to the array of excessive rotor vibration problems that can plague virtually any type of rotating machinery, centrifugal pumps also can additionally suffer from excessive vibration excited by the hydraulic instability and pressure pulsation phenomena described in the previous section of this chapter. Whether designing a new pump configuration or troubleshooting a

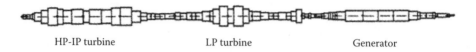

HP-IP turbine   LP turbine   Generator

FIGURE 3.6
FEA model of a main steam turbine-generator rotor.

plant operating pump in vibration trouble, use of computer vibration models is desirable. Obtaining the important vibration characteristics of a machine or structure from a large degree-of-freedom (DOF) model is not nearly as daunting as one might initially think, because of the following axiom: Rarely is it necessary in engineering vibration analyses to solve the model's governing differential equations of motion in their totality. For example, *lateral rotor vibration (LRV) analyses* generally entail no more than the following three categories.

1. Natural frequencies (damped or undamped) and corresponding mode shapes
2. Vibration amplitudes over full-speed range from rotor mass unbalances
3. Self-excited vibration threshold speeds, frequencies, and mode shapes

None of these three vibration analysis categories actually entails obtaining the general solution for the model's coupled differential equations of motion. That is, the needed computational results can be extracted from the model's equations of motion without having to obtain their general solution, as explained in Adams (2010).

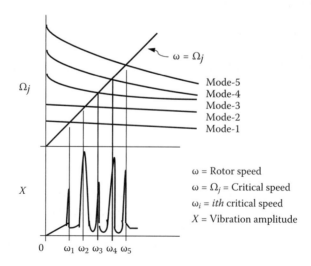

FIGURE 3.7
Campbell diagram combined with unbalance vibration amplitude.

The model for the rotor alone is assembled using a standard finite element analysis (FEA) model, shown by the main steam turbine-generator FEA rotor model in Figure 3.6.

To complete the computer model, the rotor model is "connected to the ground" with the linearized stiffness and damping coefficients for the bearings, seals, and any other interactive rotor forces (e.g., balancing drum; see Figure 2.7), support structure, and foundation. Since the interactive rotor-to-ground dynamic connection properties are usually speed-dependent, the vibration characteristics of the entire vibration model are therefore likewise speed-dependent. This is compactly summarized in Figure 3.7, which superimposes the speed-dependent natural frequencies versus rotor speed (i.e., a *Campbell diagram*) and the unbalance driven vibration amplitude versus rotor speed. The so-called *critical speeds* are the rotor speeds at which the residual rotor mass unbalances excite each natural-frequency mode to resonance.

3.2.1 Rotor Dynamical Natural Frequencies and Critical Speeds

To fully understand what Figure 3.7 compactly summarizes, one need only consider the simplest of rotor vibration models, illustrated in Figure 3.8. This is a 2-DOF system in which the rotor mass is modelled as a single mass point with only x-y planar orbital motion coordinates. The combined shaft and support structure x and y flexibilities are modeled with linearized spring stiffness and damper coefficients. For small motions, the two motion coordinates (x, y) are decoupled (by the trigonometry of very small angles), that is, do not interact, so it is like two 1-DOF systems. The rotor unbalance is modeled by its equivalent force that rotates with the rotor at speed ω. The two decoupled equations of motion are thus as follows:

$$m\ddot{x} + c_x \dot{x} + k_x x = F_o \cos \omega t$$
$$m\ddot{y} + c_y \dot{y} + k_y y = F_o \sin \omega t \quad (3.1)$$

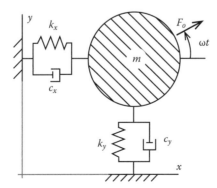

FIGURE 3.8
Simplest lateral rotor vibration model for radial-plane orbital motion.

This simple model has two natural frequencies: $\omega_x = \sqrt{k_x/m}$ and $\omega_y = \sqrt{k_y/m}$. Since the resonance in the x-direction and in the y-direction each are thus of the classical 1-DOF system, the vibration harmonically excited steady-state amplitude is given by the well-known 1-DOF steady-state harmonic response graphed in Figure 3.9.

Even in multi-DOF systems, the resonance-vibration-amplitude peaks (Figure 3.7) are like the same phenomenon as in the 1-DOF system response shown in Figure 3.9. For actual machines, either from real vibration measurements or computer models thereof, there is also typically a relatively sharp phase-angle transition similar to that for the 1-DOF model as it transitions through its natural frequency (Figure 3.9b), but typically not the 180° shift of the much simpler 1-DOF system. Interestingly, in extremely well bounced rotors, roll-up or coast-down vibration amplitude measurements may not provide adequate vibration amplitudes to nail down what are the critical speeds ω_i because of the ever-present signal *background noise*. In such cases,

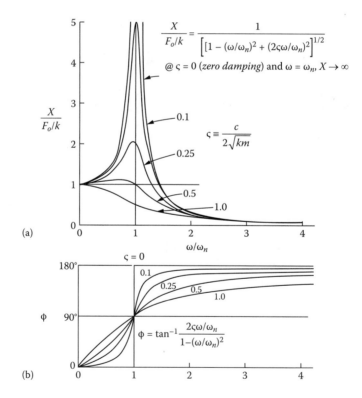

FIGURE 3.9
One-DOF steady-state response to a sinusoidal force. (a) $X/(F_o/k)$ versus ω/ω_n and (b) phase angle ϕ versus ω/ω_n, $\varsigma = 1$ (critically damped).

Operating Failure Contributors

it is usually possible to locate the critical speeds by tracking the phase-angle transition of the bandwidth filtered synchronous (once-per-rev) vibration amplitude component. The critical speeds become apparent by locating the associated relatively sharp phase-angle transitions.

3.2.2 Self-Excited Dynamic-Instability Rotor Vibrations

The equation of motion for the well-known 1-DOF model for unforced vibration is as follows:

$$m\ddot{x} + c_x \dot{x} + k_x x = 0 \qquad (3.2)$$

The range of transient responses typically presented for this system is graphed in Figure 3.10. However, mechanical structures of all kinds usually fall into the category of so-called *underdamped* systems, that is, damping ratio $\varsigma \equiv c/c_c < 1$, where $c_c = 2\sqrt{km}$ is the damping coefficient value for the *critically damped* condition. That is, the damping coefficient value that yields the fastest nonoscillatory transient motion decay.

Unfortunately most textbooks on vibration do not show the following additional simple case. If one simply inserts a negative value for the damping coefficient c into Equation 3.2 for the 1-DOF model, the motion types shown in Figure 3.10 are supplemented with the response for $c < 0$ shown in Figure 3.11. Understanding this simple example is the key to understanding the general vibration category called *self-excited vibration*.

Determination of thresholds for self-excited rotor vibration thus is one of determining the value of the controlling parameter (e.g., rotor speed, pump flow, component wear) at which an unforced-system natural-frequency mode becomes dynamically unstable. That is, determining the value of the

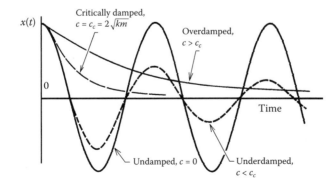

FIGURE 3.10
Motion types for the unforced 1-DOF system.

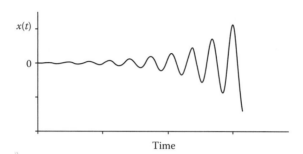

FIGURE 3.11
Initial growth of dynamical instability from an initial disturbance.

controlling parameter where the real part of one of the system equation-of-motion roots λ (eigenvalues) transitions from negative to positive. Each vibration system natural mode has its motion formulation expressible by Equation 3.3, where the real part of its two eigenvalues λ_j embodies the mode's damping and the imaginary parts $\pm i\omega$ embody the mode's harmonic (sinusoidal) contribution:

$$\{x\} = \{X\}e^{\lambda t} \tag{3.3}$$

where $\lambda = \alpha \pm i\omega_n$, ω_n = natural frequency, and $i = \sqrt{-1}$.

Clearly, if alpha > 0, the mode self-excites per Figure 3.11. If alpha = 0, the mode behaves as the undamped case in Figure 3.10. If alpha < 0, the mode behaves as the underdamped case in Figure 3.10. For a rotor vibration mode that transitions into a self-excited vibration, the typical exponential growth of the rotor orbital vibration, for example at a journal bearing, is illustrated in Figure 3.12a. The initial transient growth of the rotor orbital motion begins with a linear-system $e^{\lambda t}$ portion that is the counterpart of the 1-DOF negatively damped linear system response illustrated in Figure 3.11. This is followed in time by its high amplitude steady-state orbital nonlinear limit cycle as illustrated in Figure 3.12b, being limited in the illustrated case only by the bearing radial clearance. The high potential for intolerably high rotor vibration levels under self-excitation is quite evident from Figure 3.12b. Unlike a rotor critical speed that with sufficient bearing damping can be safely passed through when accelerating to operating speed, instability thresholds generally cannot be alleviated by increasing the controlling parameter like speed. This is presented in more depth by Adams (2010) in explaining the *hysteresis loop* for rotor-bearing self-excited orbital vibration.

Operating Failure Contributors 59

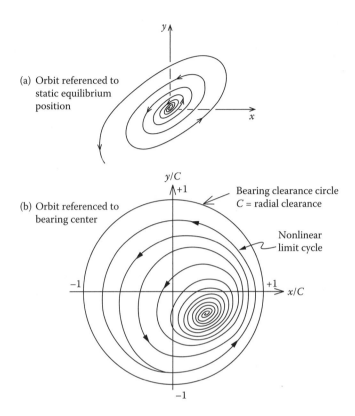

FIGURE 3.12
Transient rotor orbital vibration buildup in an unstable condition: (a) initial linear transient buildup and (b) growth to nonlinear limit cycle.

3.2.3 Dynamic Forces Acting on the Rotor

The research test results presented in Figure 1.14 provide a compact example of amplitudes for hydraulically induced dynamic rotor forces over the full pump operating flow range. Drawn from his extensive laboratory and troubleshooting field experiences, Makay (1977) assembled the all-encompassing chart in Figure 3.13 for pump dynamic rotor forces. From a rotating machinery vibration specialist's perspective, centrifugal pumps are the most challenging type of machinery. The troubleshooting case studies presented in Section III of this book show that excessive vibration and measured vibration frequency spectrum signatures are frequent symptoms indicative of a variety of pump operating problem root causes, as well as an operating problem per se.

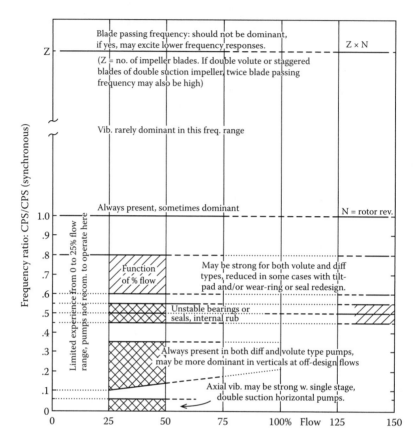

FIGURE 3.13
Frequencies of dynamic rotor forces versus operating flow range.

3.3 Wear

Like any piece of machinery, centrifugal pumps are susceptible to many different kinds of *wear*, the gradual process of material removal. Wear is an extremely complicated subject with a large variety of different fundamental mechanisms. For example, wear occurs from actions such as rubbing contact, erosion, chemical reactions like corrosion, impacting particles, and fluid dynamical impingement. The most prominent wear categories are schematically named in Figure 3.14. A long history of empirical investigations has led to insight, predictive tools, and comprehensive books to deal with the various known wear mechanisms. The *Wear Control Handbook* edited by Peterson and Winer (1980) is perhaps as complete a reference on wear as any, but there are a number of comprehensive subsequently published works on wear.

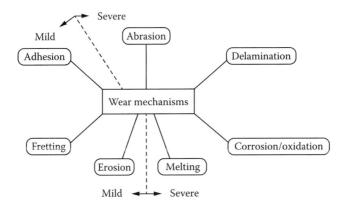

FIGURE 3.14
Most prominent mild and severe wear mechanisms.

3.3.1 Damage Caused by Pump Cavitation

In Chapter 2, Section 2.3 describes the fundamental pump cavitation phenomenon and shows a picture example of a cavitation-damaged impeller in Figure 2.3. The requirements for pump net positive suction head (NPSH) needed to prevent cavitation and the laboratory testing method to determine that are also covered in Section 2.3. The fundamental wear mechanism for cavitation comes under the category *erosion*, shown in Figure 3.14, where the cavitation damage is caused by fluid impingement from a high velocity microjet created by the sudden nonspherical collapse of the cavitation vapor bubbles. Cavitation damage is the one wear mechanism that most sets a centrifugal pump apart from other machinery types. An illustration of cavitation at an impeller vane inlet region is illustrated in Figure 3.15 (Makay 1994). On the illustrated low-pressure side of the impeller vane, the vapor cavity region of length L_{cav} and the erosion damage region of length L_{Damage} are clearly distinguishable.

Pump designers often make reference to a *material property* they call *cavitation resistance*. But there appears to be no quantitative parameter by which it can be evaluated over the range of materials suitable for pump internal components, only some heuristic relative measures. For example, as the intensity of cavitation becomes sufficient to damage an impeller, the material failure mechanism is thought to progress from a fatigue-like process over plastic deformation to a failure mechanism where the cavitation intensity exceeds the tensile strength of the material. An ultimate material resilience U_R is proposed by Hammit (1980) as the best compromise. Guelich and Pace (1986) provide guidelines that encompasses modern research findings on centrifugal pump cavitation damage prediction.

From the richly endowed published literature on wear, one can readily delve deeply into the full range of wear mechanisms named in Figure 3.14. Here a brief summary is given for some of these that are particularly relevant to wear in machinery such as centrifugal pumps.

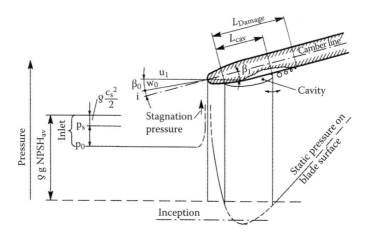

FIGURE 3.15
Static pressure distribution on low-pressure side of impeller vane.

3.3.2 Adhesive Wear

Adhesive wear is most probable in all unlubricated surface-to-surface rubbing contacts. But even in unlubricated surfaces, naturally formed oxides become a *dry lubricant*. Significant adhesive wear can also occur between lubricated rubbing surfaces, but at a reduced rate depending upon the rubbing materials' strength properties and the *boundary lubrication* capability of the intervening lubricant. The classical Coulomb friction coefficient is often approximated as a "constant" independent of the normal force magnitude that pushes together the two bodies in relative sliding motion. Figure 3.16 hypothesizes what can occur at the microscopic level between two surface asperities in rubbing contact. Two contacting asperities are postulated to momentarily form an adhesive bond called *cold welding*. In one possibility, this bond is simply broken as the two surface asperities each go unaltered their own way as illustrated. In the other possibility, the adhesive bond

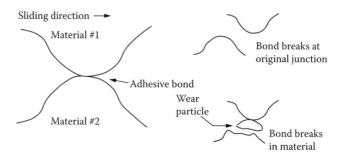

FIGURE 3.16
Adhesive material rubbing contact yields two possible phenomena.

between the two asperities overcomes the yield strength of the weaker of the two asperities, creating a wear particle as also illustrated.

When two solid bodies are pushed together even at light loads, some asperities of the two bodies come into contact with the weaker asperity deforming plastically, that is, yielding at the tip. These plastically deformed asperity tips are assumed to have a contact pressure equal to the yield strength of the weaker material. Satisfying *static equilibrium* for either body then requires the yield strength of the weaker material times the combined contact area of all the asperity flattened tips must equilibrate the total compressive force that is pushing the two bodies together. The force needed to perpetuate sliding will equal the sum of sliding friction forces from all of the asperities in contact. Accepting that the weaker material's yield strength does not change in the process, the combined surface area of all the flattened asperities will increase in proportion to the force pushing the two bodies together. So the force required to slide the one body with respect to the other will be approximately in proportion to the force pushing the two bodies together. That naturally means the proportionality coefficient relating the friction force to the compressive force is a constant, that is, the Coulomb friction coefficient. This is, of course, a useful simplification of a very complex phenomenon. In fact, modern, closely controlled laboratory experiments on rubbing friction have shown that with a fixed normal force, the sliding friction force varies somewhat randomly with the sliding velocity magnitude, but typically remains significantly smaller than the breakaway static friction force needed to start sliding. To better approximate this for engineering purposes, two values for the friction coefficient are typically used, namely, the *static* value μ_S and *dynamic* value μ_D, where $\mu_S > \mu_D$.

Referring again to Figure 3.16, it illustrates the release of a wear debris particle when an asperity tip breaks off. *Archard's law* is the longstanding approach for predicting wear volume or rate of volumetric wear derived from the concept illustrated in Figure 3.16. It is expressible in many alternate forms including Equation 3.4. Predicating a probability ($0 < k < 1$) for a contacting asperity tip to break loose, the volume of wear debris for a sliding event is predicted. V is the wear volume, P is the applied normal force, x is the sliding distance, v is the sliding velocity, and H is the material hardness. The number 3 occurs in Equation 3.4 because the derivation includes the formula for assumed *hemispherical* volume of wear particles (Volume = $2\pi r^3/3$).

$$V = \frac{kPx}{3H} \text{ and } \frac{dV}{dt} = \frac{kPv}{3H} \qquad (3.4)$$

The factor of 3 is of small consequence because the wear coefficient k varies over several orders of magnitude. This wear law was developed before friction-and-wear researchers had access to electron microscopes with which they subsequently discovered that "adhesive" wear particles are anything but hemispherical (see Section 3.3.4). Based on extensive laboratory test results from many sources, the wear coefficient k ranges from 1 to 1500 times 10^{-6} with V in cubic meters (m³), depending upon the effectiveness of any lubrication present and the degree of compatibility between the two rubbing materials.

3.3.3 Abrasive Wear

Abrasive wear occurs when particles of a softer material are rubbed off by a harder rough surface. The classic example is sandpaper rubbed against wood. Abrasive wear also can occur when a hard third substance gets into the rubbing contact area, like when machinery is operated in a desert environment or an abrasive is intentionally injected like in lapping and polishing processes. An abrasive can also be formed from the rubbing surfaces such in *fretting* where small oscillatory rubbing produces wear particles that cannot escape the contact region and subsequently oxidize into an abrasive third body that is harder than the rubbing surface materials. One approach used for predicting abrasive wear is to use the same probability-based approach implicit in Archard's law for adhesive wear. This is given in Equation 3.5. The ranges for k_{abr} are 10^{-4} to 10^{-3} for strong abrasion and 10^{-7} to 10^{-6} for mild abrasion.

$$V = k_{abr}\frac{Px}{H} \text{ and } \frac{dV}{dt} = k_{abr}\frac{Pv}{H} \qquad (3.5)$$

3.3.4 Delamination Wear

Research by Suh (1977) and colleagues at the Massachusetts Institute of Technology (MIT) in the 1970s significantly advanced the understanding of the mechanics for rubbing adhesive-like wear. Using an electron microscope, they discovered that the wear particles liberated through the assumed adhesive action are in fact not hemispherical but are flat sheets that delaminate from the contact. The fundamental view of the *delamination wear theory* is the formation of subsurface microcracks that propagate in time as a *fracture mechanics* phenomenon until liberated when acquiring some critical length. This is reminiscent of the longstanding experimentally validated mechanics basis for life rating rolling-element bearings, albeit that is based on subsurface initiated fatigue cracks caused by traveling intermittent Hertzian contacts that lead to *raceway spalling* (see Chapter 2, Section 2.4.3).

3.4 Operating Problem Modes

Many of the large fossil-fired units still in service were initially commissioned as base-load units operating at near-rated capacity most of the time.

Operating Failure Contributors

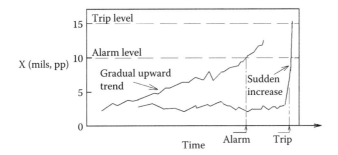

FIGURE 3.17
Tracking of two representative vibration amplitudes over time.

But the emergence of nuclear power plants, which are optimally profitable only at rated capacity, necessitated that many of the prior base-load fossil units be continuously cycled for load following. That produced unanticipated operating distresses on a whole host of these fossil plant systems, especially including the major pumps.

Plant operators need to understand the fundamental differences between a sudden pump component failure or sudden pump failure-to-operate and the more gradual pump degradation scenarios. The topics covered in the prior sections of this chapter provide a background to better understand the causes of both *sudden failures* and degradation *gradually approached failures*. The two vibration situations illustrated in Figure 3.17 are obvious symptomatic contrasting examples of these two extreme situations that both can be labeled symptoms of a machine malfunction. In one case it is apparent that some degradation is progressing slowly but surely to the point where the machine will have to be taken out of service to identify and rectify the malfunction route cause. In the other case the failure has occurred almost instantaneously without warning. Florjancic (2008) lists a wealth of pump malfunction root causes, identified and included in the following sections.

3.4.1 Rotor Mass Unbalance

The pump shaft holds the impeller(s) and other rotating parts such as the shaft seal and thrust balancer components, thrust bearing runner, and shaft coupling. In a rigidly coupled pump-driver assembly, the "rotor" is the completely assembled pump-driver combination. Rotor mass unbalance is a major source of excessive once-per-revolution (synchronous) vibration amplitudes. In particular, an improperly shop dynamic rotor balancing of individual rotating parts as well as the completely assembled rotor and poorly executed erection are suspect rotor unbalance root causes when the pump is first operated. In service, rotor unbalance can accrue from material losses initiated by wear phenomena including cavitation, corrosion, and abrasion. Deposits of foreign material on the impeller(s) can also progressively degrade the balance quality of the rotor. Inaccurate dimensions of impeller cast channels can also be a source of once-per-revolution vibration due to the resulting nonsymmetric flow distortions within the impeller. This understandably is misdiagnosed as rotor mass unbalance. It is difficult to alleviate with impeller weight-distribution corrections because it varies with pump operating flow. That is, it is not as simple a phenomenon as the center-of-gravity of the liquid within the impeller being off center.

3.4.2 Unfavorable Rotor Dynamic Characteristics

As mentioned in Section 3.2, it is desirable to analyze, with appropriate computer modeling, the rotor dynamic characteristics of a pump at the design stage, since some rotor vibration problems can already be eliminated at that point. For example, ensure the absence of any critical speeds within the continuously operating speed range. Also, avoidance of self-excited rotor vibration as identified in Section 3.2.2 can be properly analyzed using computer models. Sources of dynamic destabilizing effects from journal bearings and other radial-gap liquid annuli can now be analyzed with the available dynamic system-identification data (e.g., stiffness and damping coefficients) on bearings and seals from research sponsored by NASA and the Electric Power Research Institute (EPRI) (e.g., Childs 1993 and Adams 2010). Radial clearances in fluid annuli (e.g., at wear rings) that become too enlarged as a consequence of accumulated wear can cause a pump to migrate into a state of excessive vibration.

3.4.3 Hydraulic Forces at Off-Design Operating Conditions

The *hydraulic instability, pressure pulsation,* and *minimum recirculation flow* topics covered in Section 3.1 are at the heart of the fluid dynamical phenomena that can create destructive level dynamic forces within a centrifugal pump. For an 85% peak-efficiency pump of several thousand horsepower, where does the other 15% go? Most of it is dissipated into fluid heating. However, the portion of this energy needed to sustain destructively high levels of vibration is relatively quite small compared to the dissipated fluid heating, potentially resulting in component failure from high-cycle fatigue. So the amount of exposure time to off-design operation is a critical factor in assessing failures or malfunctions. Plants devoted to load following are consequently at higher risk for pump failure than those for base load units.

One of the major shortcomings that Makay (1977–1994) exposed in his pioneering troubleshooting work is the overriding emphasis that the pump designers had placed on peak efficiency in response to the high priority given to peak efficiency by plant engineers' pump purchase specifications. The design consequence of this led to improper sizing of the radial gaps between the impeller and the stationary casing. Figure 3.18 illustrates these critical dimensions, Gap A and Gap B as named by Makay. By dimensioning these

Operating Failure Contributors 67

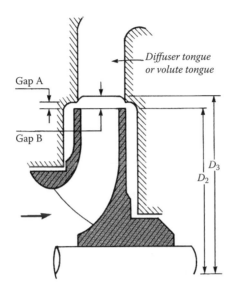

FIGURE 3.18
Gap A and Gap B between impeller periphery and stationary vanes.

gaps to maximize peak efficiency at the BEP, the efficiency curve overall is relatively peaky. Thus at off-design operating flows, the pump efficiency suffers somewhat. This means the flow is quite "unhappy" at off-design flows and the accompanying unsteady flow is excessively violent.

As these fossil units were subsequently transitioned from base load to load following duty, off-design operating time durations significantly increased. Figure 3.19 clearly illustrates the violent flow-erosion damage of a stationary vane resulting from extensive low flow pump operation during load cycling. Figure 3.20 illustrates a material fatigue damaged impeller sidewall (shroud) resulting from the same violent off-design unsteady flow accentuated by improperly sized Gap A and Gap B dimensions.

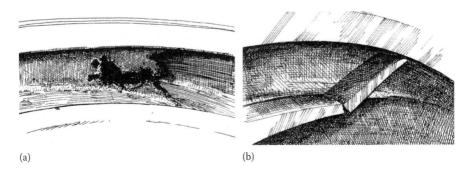

(a)   (b)

FIGURE 3.19
(a) Eroded diffuser vane and (b) diffuser vane after piece breaks.

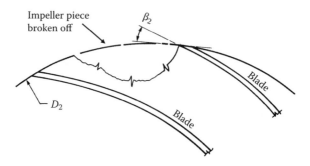

FIGURE 3.20
Impeller fatigue failure from high-amplitude pressure pulsations.

Makay's solution to this industry-wide problem was simply to properly size Gaps A and B with minimal impact on the BEP efficiency, but yielding overall a less peaky efficiency versus flow curve. From numerous power plant field measurements, Makay (1977–1994) developed the pressure pulsation intensity curve shown in Figure 3.21. The dimensionless pressure pulsation force F is normalized by that at a radial Gap B of 4% of the impeller radius $D_2/2$. Most important, Makay's pressure pulsation intensity curve shows the steep increase in pressure pulsation force as Gap B is progressively reduced. His fix was to standardize a 4% radial gap as the minimum recommended.

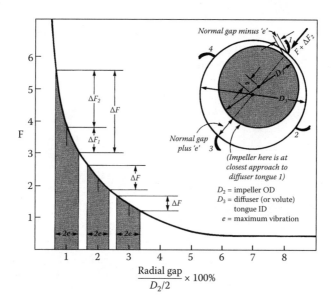

FIGURE 3.21
Pressure pulsations relative magnitude versus impeller radial Gap B.

3.4.4 Dynamic Characteristics of Foundation, Support Structure, and Piping

In a number of troubleshooting experiences, the elusive root cause of plant rotating machinery vibration problems has been determined to be the neglect at the design stage that the support structure is not immoveable. Specifically, suppose a machine is to be mounted to the floor of a large plant and the vibration design analysis of the machine assumes the floor to be perfectly rigid. The vibration analysis will likely be seriously flawed. Common sense dictates that one does not devise an FEA model of the entire plant building just to couple it to the vibration model for the pump alone. Usually it is the vertical floor motion under the machine that is the important feature that needs to be incorporated into the overall vibration analysis model. The approach for characterizing any linear electrical circuit's impedance as a simple LCR circuit is by imposing a *controlled harmonic voltage* and measuring the resulting *harmonic current output* over the applicable frequency range. By varying the frequency of the imposed voltage, equivalent LCR coefficients can be solved as functions of frequency. For the vibration counterpart, a controlled harmonic vertical force is imposed upon the floor and the resulting floor vibration simultaneously measured. Figure 3.22 shows a schematic of this and Equation 3.6 is the system equation of motion (Adams 2010):

$$(m_s + m_f)\ddot{x}_f + c_f\dot{x}_f + k_f x_f = F_s e^{i\omega t}$$
$$x_f = Xe^{i(\omega t + \phi)}$$
(3.6)

Equation 3.6 leads to the following complex algebraic equation:

$$(k_f - \omega^2 m_f - \omega^2 m_s + ic_f\omega)Xe^{i\phi} = F_s \quad (3.7)$$

Equation 3.7 is equivalent to two real equations and thus can yield solutions for two unknowns—$(k_f - \omega^2 m_f)$ and c_f—at a given frequency. These shaker test results are then used to "connect" the pump dynamics model to the ground (inertial reference frame).

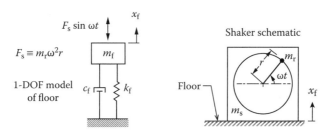

FIGURE 3.22
Vertical shaker test of a floor where a machine is to be installed.

Hydraulic instability flow surging can significantly excite excessive vibration of the whole piping system connected to the pump. This is an unfortunate consequence of a piping system's natural resonance frequency fortuitously aligning close to one of the pump's strong forcing frequencies. Although this might not be viewed by the pump supplier as the pump's fault, keeping the customer happy makes it advisable for the supplier to identify and help alleviate the problem.

3.4.5 Unfavorable Pump Inlet Flow Conditions

Inlet flow conditions have pronounced influences on centrifugal pump behavior, especially for single stage pumps. Low inlet pressure resulting in insufficient NPSH induces cavitation, which not only can lead to accrued damage to the impeller vanes but also a degradation of pump steady flow and efficiency. Unfavorable suction piping design, especially for double-bend piping in different planes, creates inlet flow distortions that degrade pump steady flow and efficiency. Similar performance degradations are also induced from the creation of swirls on the surface of the sump of vertical pumps. The higher the specific speed of a pump, the greater the degrading influences of nonuniform impeller inlet velocity distributions. It follows from the inlet velocity triangle (Chapter 1, Figure 1.8), and the associated derivation leading to Equation 1.6 (Chapter 1), that impeller corotational inlet preswirl reduces pump head produced, whereas counterrotational inlet preswirl increases pump head produced.

3.4.6 Bearing, Seal, Shaft, and Thrust Balancer Damage

The relationship between bearings, seals, and the shaft is easily explainable even to grade-schoolers. The pump *shaft* carries all the rotating parts combined into what is called the *rotor*. The *bearings* keep the spinning rotor positioned where it needs to be. The *seals* keep the pumped liquid inside the pump and can keep out external contaminants from getting inside the pump. If any one of these parts stops doing their job, the pump stops doing its job. Section 2.4 in Chapter 2 describes these parts.

For small single-stage pumps, rolling-element bearings are typically used. For larger pumps (e.g., multistage), fluid-film bearings are generally used. Rolling-element bearings are all life rated based on subsurface initiated fatigue (spalling) as explained in Section 2.4. But they are vulnerable to other failure mechanisms that can occur sooner than the rated life, for example, improper installment setting, inadequate lubrication, hard particle dirt ingestion, moisture, and corrosion. A rolling-element bearing that is in distress often gives an audible warning by radiating a winning sound. In recent times, researchers have developed model-based schemes utilizing vibration and acoustic measurements, nonlinear modeling, and advanced data analysis methods to detect impending bearing failures before they occur (e.g., Adams and Loparo 2004).

Fluid-film bearings running normally have a small but finite lubricant film thickness. Without metal-to-metal sliding contact, they are theorized as being able to last indefinitely. Of course, reality is invariably to the contrary. For example, influences that limit the life of fluid-film bearings include starved lubrication, insufficiently filtered-out hard dirt particles in the lubricant, excessive rotor vibration levels (fatigue), elevated operating temperatures such as from gross overload, design shortcomings, misalignment, and flawed assembly. The major monitored fluid film bearing signals that announce bearing distress are from embedded thermal couples to monitor bearing surface temperatures. Florjancic (2008) includes an extensive group of photographs showing damaged bearings with the various root causes identified.

Shaft breakage is among the most dreaded of failure scenarios because of the added consequential destruction of rotating and nonrotating internal components. As covered in any undergraduate text on machine design, the static radial shaft loads produce oscillating fatigue-initiating beam bending stresses in the shaft. That is, the beam bending stress field is not rotating but the shaft is. Section 2.4.1 with Figure 2.8 (Chapter 2) describes shaft failure scenarios for multistage pumps. A crack propagating through a spinning shaft produces an emerging nonaxisymmetric bending stiffness of the rotating shaft, that is, two different principle bending moments-of-inertia. This produces two emerging effects to the monitored rotor x-y orbital vibration signals. First is the gradual emergence of a twice-running-speed (2N) frequency component, since the shaft now has emerging a *maximum* and a *minimum static bending sag line*. Second is the gradual emergence of a crack-local shift in the bending neutral axis, causing a localized rotor bow corresponding to the crack direction. This second effect transiently adds vectorially to the preexisting residual unbalance synchronous vibration. The gradual emergence of these two simultaneous rotor vibration symptoms is now widely employed successfully to predict operating time remaining before the shaft breaks (Adams 2010). More recent research (e.g., Laberge 2009; Laberge and Adams 2007) provides a means to further detect the progress of a shaft crack and its axial location utilizing the crack-closing stress-wave propagation intensity and speed.

For optimum power efficiency, the more likely configuration now for multistage pumps is to have all the stages pointing in the same direction. As explained in Chapter 2, Section 2.4.5, high-pressure multistage pumps then typically can have very high hydraulic axial forces that load the rotor well beyond the load capacity of any conventional thrust bearing. Thus failure of the axial balancing device in these pumps is a dreaded failure scenario leading also to the added destruction of other rotating and nonrotating internal pump components. Thrust balancers are either of the balancing drum or balancing disk type, illustrated in Figures 2.28 and 2.29. In Makay's (1978) survey of boiler feed pump outages in many plants, 533 were equipped with balance disks with 310 reporting failures. In 511 pumps equipped with balance drums, there were 27 reported failures. The apparent inferiority of the balance disk configuration was attributed both to initial design under sizing of the balance disk and also heat treating them to too high a hardness that increased their propensity to fail from cracking.

3.5 Condition Monitoring and Diagnostics

The operability of most major systems in a power plant is dependent upon operability of pumps. A number of measureable operating parameters have been found usefully informative in evaluating the operating health of centrifugal pumps. The cost effectiveness for monitoring each specific operating parameter must be assessed based on factors such as

- Initial cost
- Suitability for monitoring continuously
- Early failure-detection indicator
- Control-room readily useable
- Troubleshooting usefulness

The most monitored machinery operating parameter is perhaps vibration, especially in power plants. On machinery critical to plant availability, vibration signals are continuously or regularly monitored and recorded, with selected continuously updated portions displayed in the control room or assessed by automated plant control. The straightforward examples in Figure 3.17 of Section 3.4 illustrate two obviously important uses of monitored vibration levels. Transforming time-base vibration signals into the frequency domain (FFT) has become a regular tool for assessing vibration and general root cause identification. Adams (2010) devotes three entire chapters to cover (1) sensors, signal acquisition, and analysis; (2) vibration severity guidelines; and (3) root cause identification. In addition to vibration monitoring, critical temperatures and pressures are continuously or regularly measured operating parameters in power plant rotating machinery, especially in the major pumps. Many large power pumps have been updated with retrofitted vibration and pressure sensors to continuously measure operating parameters that through extensive troubleshooting experiences have been found essential in assessing pump health and providing forewarning of potential pump failures, as shown in Figure 3.23.

Operating Failure Contributors

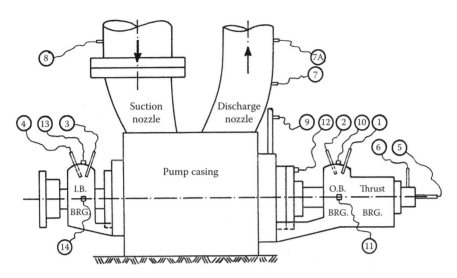

FIGURE 3.23
Multistage high power pump time-base condition monitoring measurements. Channels 1 to 6 are shaft-relative-to-bearing vibration displacement proximity probes, channels 7 to 9 are pressure transducers, and channels 10 to 14 are accelerometers.

3.5.1 Vibration Measurement

As any engineering undergraduate knows, the three elementary dynamics/ kinematics parameters are *displacement*, *velocity*, and *acceleration*, directly relatable to one another through elementary calculus. And these are the three vibration parameters that are regularly measured and monitored on plant operating machinery. The sensors for these three vibration signals are schematically illustrated in Figure 3.24. For accelerometers, the accurate useable frequency range must be considerably below the accelerometer's own 1-DOF natural frequency. For a velocity sensor, the useable frequency range must be sufficiently above the sensor's own 1-DOF natural frequency. Because velocity sensors are intrinsically fragile, they are primarily for laboratory use. So an industrial velocity sensor is in fact an accelerometer with a built-in signal integrator. For noncontacting position sensing, the inductance proximity probe must be selected for the shaft material. Because of residual magnetism in a shaft, there will be some indicated runout that is not mechanical. So in extra high-accuracy applications, the rotating target is chrome plated to filter out the electrical run out (Horattas et al. 1997).

Adams (2010) gives detailed technical information for selecting sensor specifications as well as the industry experience-based vibration severity guideline shown in Figure 3.25. Although both parts of this guideline essentially contain the same guideline information, one clearly sees the appeal of using the *velocity severity levels* since a particular velocity peak value has the

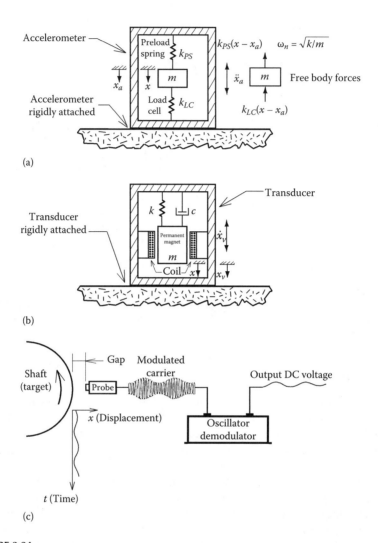

FIGURE 3.24
Vibration measurement sensors. (a) Accelerometer, (b) velocity transducer, and (c) inductance-eddy-current noncontacting position sensing.

same severity interpretation over the entire frequency range of concern for most plant machinery. A *vibration displacement severity criteria* utilizing journal position relative to journal bearing, as measured by proximity probes, is provided by Eshleman (1999) (see Table 3.1).

In analyzing any measured time base signature like vibration, digitally transforming from the time domain into the frequency domain (FFT) is now one of the most-frequently used signal analysis methods. Figure 3.26 provides a visually insightful picture of what the FFT mathematical transformation operation does.

Operating Failure Contributors

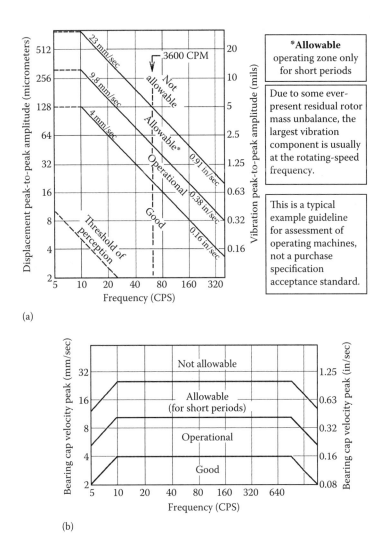

FIGURE 3.25
Bearing cap vibration acceleration and velocity amplitude guidelines: (a) displacement p-p, (b) velocity v-p; (a) and (b) contain the same information.

TABLE 3.1

Journal Vibration Guideline for Displacement with Respect to Bearing

Speed	Normal	Surveillance	Plan Shutdown	Immediate Shutdown
3600 rpm	R/C < 0.3	0.3 < R/C < 0.5	0.5 < R/C < 0.7	R/C > 0.7
10,000 rpm	R/C < 0.2	0.2 < R/C < 0.4	0.4 < R/C < 0.6	R/C > 0.6

Notes: R, peak-to-peak journal-to-bearing displacement; C, diameter bearing clearance.

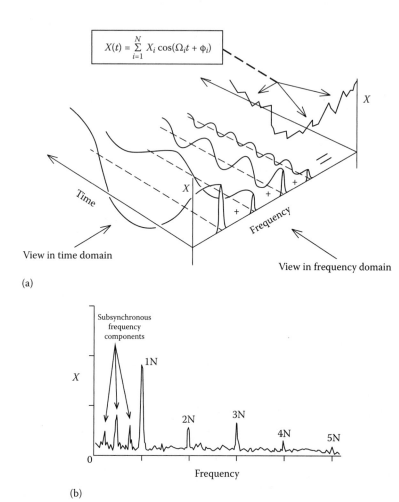

FIGURE 3.26
Fast Fourier transform (FFT) of time base signals: (a) general example and (b) rotor vibration example.

3.5.2 Pressure Pulsation Measurement

The importance of pressure pulsation measurements is twofold. First, it is a good relative measure of pump hydraulic discomfort as the operation further intrudes into off-design operating flows. Second, it is also a good relative measure of the life-shorting abuse to pump internals by pump hydraulic discomfort. Figure 3.23 shows pressure transducer locations on a large multistage pump (Makay 1977–1994). He supplements that with a troubleshooting experience-based insightful illustration of pressure pulsation measurement from *balance-disk leak-off flow* (Figure 3.27). These monitored pressure pulsations are indicative of the low-flow guideline in Figure 3.5. Makay also

Operating Failure Contributors

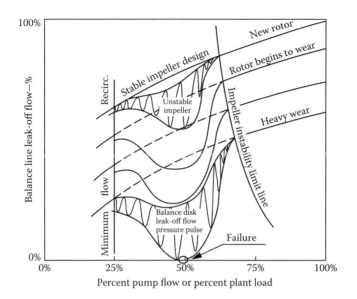

FIGURE 3.27
Pressure pulsations warn of pump discomfort/damage at low flows.

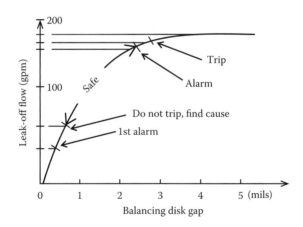

FIGURE 3.28
Balancing-disk leak-off flow as an indicator of feed pump health.

highly recommends that balance-disk leak-off flow should be monitored at all times for pump health assessment, as illustrated in Figure 3.28.

3.5.3 Temperature Measurement

Temperature measurements are relatively low first cost and are made with embedded thermal couples. Bearing temperatures are a reliable measure of bearing operating health. Pump suction and discharge temperatures are reliable indicators of pump internal health. Depending upon the specific configuration and application of a pump, combinations of temperatures along with other monitored parameters taken together can reliably provide early warning of impending operating problems.

3.5.4 Cavitation Noise Measurement

The sound intensity of cavitation noise in centrifugal pumps has been investigated by several researchers. The objective of that research has been to develop a reasonable design engineering approximation to predict the amount of cavitation-caused impeller damage as a function of operating time and pump inlet conditions over the used part of the pump's *H-Q* curve. Piezoelectric pressure transducers with very high frequency response, way beyond auditory limits, have been employed in pump controlled testing. A consensus among pump hydraulic specialists is that the cavitation phenomenon does not lend itself to being a reliable cavitation-damage predictive tool based upon measureable noise. Using a specific pump installed into a specific loop, it is reasonable to anticipate experimentally repeatable correlations between cavitation noise signatures and cavitation-caused impeller damage. However, such a repeatable correlation is likely to be transferrable to neither a different pump nor the same pump in a different loop. That means a power plant operating pump would have to serve as the research test setup. Naturally one can forget that because power plant owners have long resented OEMs using plant installed pumps as de facto test rigs to complete their research.

3.5.5 Pump Test Rig for Model-Based Condition Monitoring

Figure 3.29 is a photograph of the Case Western Reserve University (CWRU) multistage centrifugal pump test loop for research in *model-based conditioning monitoring* for power plant pumps (Adams 2016). Its development is in response to pump condition monitoring in general, but specifically in response to plant services where the pumps are submerged and thus impractical for periodic condition inspections, for example, river pumps for nuclear power plants. Figure 3.30 shows the submerged accelerometer locations on the CWRU test pump inner casing surface. Figure 3.31 shows sensor locations on the test pump as it is submerged in the transparent test-loop outer can. A fairly new player in pump condition monitoring is the Robertson efficiency probes (Robertson and Baird 2015). A matched pair of these probes accurately measures pump efficiency in real time using ultra-precise temperature difference measurements (Figure 3.32). Pump *efficiency degradations* can be a valuable complementary parameter in detecting pump deteriorating health and thereby giving faulty detection warnings to avoid forced outages.

FIGURE 3.29
(See color insert.) CWRU multistage centrifugal pump research test loop.

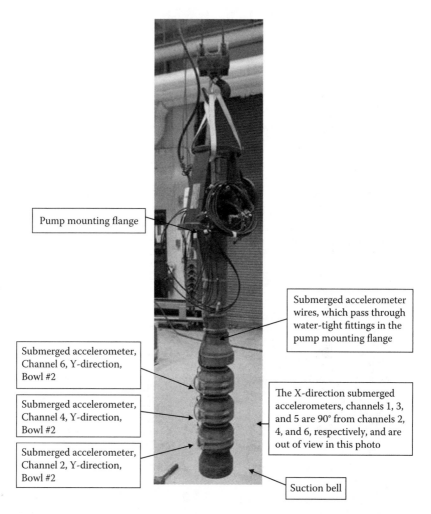

FIGURE 3.30
Submerged accelerometer locations on CWRU three-stage test pump.

The fundamental premise of model-based condition monitoring is to reconstruct in real time the behavior conditions inside the pump from a computer model driven by measured signals from external readily accessible sensor locations on the pump-driver unit. The CWRU test facility shown in Figures 3.29 through 3.32 is configured with sensors inside the unit, where sensors are not feasible to normally place in power plant operating pumps. And additionally, sensors are located at readily accessible external locations. In this research, competing real-time computer models are "tested" to determine their adequacy in replicating the internal measurements from only the signals of the external sensors. This ongoing research is quite likely

Operating Failure Contributors 81

FIGURE 3.31
Sensor locations on installed CWRU three-stage test pump.

FIGURE 3.32
Robertson pump-efficiency probes.

TABLE 3.2

Recommended Monitoring for High-Speed and Multistage Pumps

Recommended for Recording	An Indication of
Pump flow, pump speed	Internal wear
Shaft vibration (amplitude vs. frequency)	Internal wear of feed pump
Balance water flow	Drum piston clearance wear
Thrust bearing temperature	Change in axial thrust[a]
Suction pressure and temperature	Cavitation[a]
Discharge pressure	Internal wear
Leak-off flow	Overheating of pump
Radial bearing temperature	Overload/wear[a]
Seal drain temperature	Breakdown of seal
Barrel temperature (top and bottom)	Casing distortion, insulation

[a] Recommended for small and simple pumps.

to ultimately benefit many other types of power plant machinery and systems as well as other industries like chemical process plants and naval drive systems.

3.5.6 Summary of Monitoring and Diagnostics

Table 3.2 summarizes recommended pump parameters to monitor and record (Florjancic 2008). Depending upon the specific configuration and application of a pump, combinations of monitored parameters can further provide early warning of emerging operating problems.

Section II

Power Plant Centrifugal Pump Applications

4

Pumping in Fossil Plants

The most basic cycle for a steam power plant is open loop and includes a steam engine (turbine), a boiler, and a means of getting pressurized water into the boiler. This is improved when closing the water loop by having a condenser for the turbine exhaust, increasing the pressure drop across the turbine to use more of the steam energy. The condenser returns the condensed feed water back to the boiler feed water pump to complete the closed loop. This cycle is additionally improved by heating the feed water with steam extracted from intermediate points of the steam turbine. This provides improvement to the cycle efficiency and deaeration of the feed water, and eliminates cold water injection into the boiler, thus eliminating the associated thermal strains on the boiler (Karassik and Carter 1960). This cycle requires a minimum of three pumps—feed water, condenser, and condenser circulating—as illustrated in Figure 4.1.

In addition to the essential pumps referenced in Figure 4.1, a number of auxiliary pumps are employed including various circulating pumps, sluicing pumps, cooling pumps, heater drain pumps, and boiler circulating pumps, each with specific problems of their own. In the early 1960s the ongoing trend to higher steam turbine inlet pressures led to the plants operating at super critical steam pressures up to 5500 psi. Karassik and Carter further describe the utilization of multiple steam turbine extraction points for heaters of the feed water and the associated multitude of heaters and heater drain pumps (see Figures 4.4 and 4.5). The aggregate of centrifugal pump functions for the steam power of propulsion drives for large marine vessels are virtually the same as land-based power plants, albeit of lower power capacity, but with even more stringent reliability and physical size requirements.

4.1 Boiler Feed Water

Boiler feed water pumps are without a doubt high-technology machines, although for many years they were more often thought of as the product of engineering art rather than engineering science. The only centrifugal pumps ever made that had a higher horsepower per unit of cubic volume of internal fluid passages are the space shuttle liquid oxygen and liquid hydrogen main

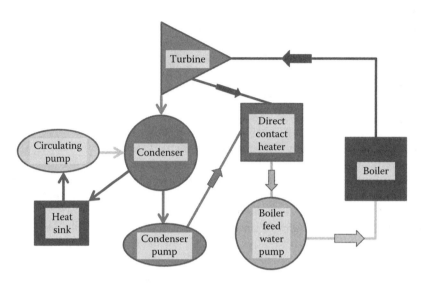

FIGURE 4.1
(See color insert.) Elementary steam power unit cycle.

engine pumps with a use time between rebuilds of less than 10 minutes, whereas boiler feed water pumps are expected to have 40,000 hours of use time between rebuilds. One is free to opine which of these two centrifugal pump applications is the higher technology one.

The typical boiler feed arrangement in present-day large fossil units is two close-to-identical boiler feed pumps in parallel to equally share the flow to the boiler, that is, two 50% pumps, sometimes with a spare change-out 50% feed water pump in reserve. With the two-50%-pumps arrangement, the generating unit can operate at approximately 65% power capacity with only one 50% pump operational, as shown in Figure 4.2

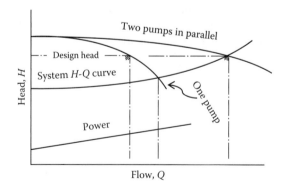

FIGURE 4.2
H-Q curves for two 50% pumps in parallel.

where the one-pump-only H-Q curve intersects the system H-Q curve. In smaller units such as industrial power plants, other feed water arrangements are also used, such as one feed water pump or several feed water pumps in parallel. When more than one feed pump is employed, they must all have stable head–capacity curves with equal shut-off heads (see Chapter 3, Section 3.1.1), otherwise they will not equally share the flow to the boiler. The large power-barrel-type boiler feed pump exampled in Figure 1.6 (see Chapter 1) has long been the workhorse of modern fossil plants. In many plants a *booster pump* in inserted just upstream of the main feed water pump to provide sufficient net positive suction head (NPSH) to the main feed pump. An alternative to a booster pump is to have the main feed water pump configured with a *double-suction first-stage pump*. Florjancic (1983) provides a comprehensive chronology of the considerable evolution of feed pump development/design dictated by the considerable increase in power ratings from 1950 to 1980 of then newly installed steam power plant units.

Current-era recently installed steam powered main turbines in the United States are now essentially all part of *combined-cycle* plants. For the combined-cycle-plant steam turbines, the feed water pumps generally adopted are of the radially split ring-section configuration such as shown in Chapter 1, Figure 1.16. It is less first-cost expensive than the superior more robust barrel-type feed water pump. If the main feed pump is variable-speed driven, care must be taken to ensure the required NPSH is met over the entire operating speed range. In large fossil plants originally commissioned as base load units, their transition to load following with the emergence of nuclear powered plants has added major stressors to the boiler feed pumps as well as to other power cycle systems.

In response to the host of resulting load-following boiler feed pump operating problems widely publicized (Makay 1978), the Electric Power Research Institute (EPRI) funded in the mid-1980s a pioneering $10 million research project to advance the robustness and reliability of large boiler feed pumps. This research involved participation of both U.S. and European major manufacturers and technologists for centrifugal pumps. The products of this research are documented in several EPRI-published interim reports and summarized in the final EPRI project report (Guelich et al. 1993). Feed pump manufacturers worldwide now have used the results of this research to improve their centrifugal pump products. The chapters of this EPRI final report address the following topics:

1. Hydraulic instability and part load phenomena
2. Cavitation erosion in centrifugal pumps
3. Rotor dynamic modeling and testing of boiler feed pumps
4. Hydraulic and mechanical interaction in feed pump systems

5. Rotor dynamic and thermal deformation test with high-speed feed pump
6. Suction system effects on feed pump performance

In a much earlier EPRI report, Makay and Szamody (1980) give the following recommended short list for operational questions that needed to be specified:

1. Pump-turbine warm up (variable speed pumps only)
2. Startup procedures
3. Switchover load point: to high pressure admission steam
4. Switchover load point: from high pressure steam admission to crossover steam
5. Manual/automatic operating modes
6. Load for placing second boiler feed pump in service
7. Control room or automatic monitoring parameters
8. Load change rates
9. Minimum flow as percent of valve-wide-open; modulating or on-off operation

Serious degradations and outright failures of boiler feed pump components, from the constant operating stressors, are not uncommon even with the best of design practices. Shaft seals, interstage sealing clearances, shafts, bearings, impellers, and axial thrust balancers are the most likely failure sources. Chapter 3 provides a primer on the fundamental failure mechanisms and root causes for component degradations and failures.

Extensive preference for the radially-split ring-section feed pump configuration (Figure 1.16) for the newer combined-cycle-plant steam turbines has apparently led to a new generation of feed pump reliability and availability problems. Were he still alive, Dr. Makay would most assuredly be extending his pioneering troubleshooting work to fix these pumps. The Architecture Engineer (AE) switch from the more robust Makay-perfected generation of the barrel-type feed water pumps are apparently motivated by lower first cost. The criticality of feed pump internal component manufacturing tolerance stack-ups, both radially and axially, were well exposed in Makay's many lucid articles and short-course notes. Just from the picture of the typical radially split ring-section feed pump in Figure 1.16, one can well imagine how their radial and axial manufacturing tolerance stack-ups can adversely impact internal flow patterns and hydraulically generated dynamic forces, as covered in Chapter 3.

4.2 Condensate, Heater Drain, and Condenser Circulating

The condensate pump suction is fed from the condenser hot well. Figure 4.1 shows where it fits into the basic steam power cycle. Both horizontal and vertical centerline condensate pumps have been employed. But large modern plants typically use vertical canned-inlet multistage condensate pumps, commonly called turbine pumps (Figure 4.3). The available NPSH is quite low (approximately 2 to 4 feet) being the difference between the water level in the condenser hot well and the condensate pump first-stage impeller. So the major advantage (i.e., NPSH) of the vertical canned-inlet configuration is that the first stage impeller is located at a lower level than with a horizontal pump. A condensate booster pump is also employed in some plants.

Controlling the condensate pump as the generating load is cycled can be accomplished with a variety of different methods depending upon the configuration of feed water heaters used. Two typical multiheater steam loops, so-called closed loop and open loop configurations, are shown in Figures 4.4 and 4.5. As illustrated, the difference between these two loops is only the use or nonuse of a direct-contact heater in-series with the closed feed water heaters. For the closed loop arrangement in Figure 4.4, the condensate pump and

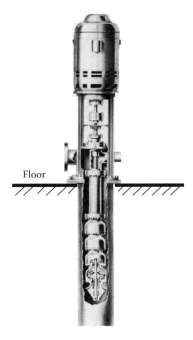

FIGURE 4.3
Vertical canned-inlet multistage vertical condensate pump (cutaway).

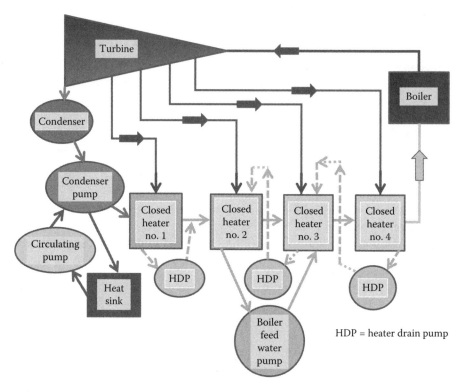

FIGURE 4.4
(See color insert.) Closed loop feed water cycle with several heaters.

boiler feed pump are like a single unit with combined single head–capacity curve. This combined head–capacity curve intersects the system head–capacity curve so flow is varied either by a throttle valve or by varying the speed of the boiler feed pump.

For an open loop arrangement (Figure 4.5), several options are in use to control the condensate pump flow:

1. *Submergence control*, allowing the pump to operate at the cavitation break point, with design measures to ensure against erosion damage, has been often successfully employed in the past
2. Throttle the pump discharge
3. Allow the pump to intersect its system H-Q curve and bypass the excess condensate back to the condenser hot well

A combination of methods 2 and 3 has also been successfully employed. However, since the submergence control approach operates continuously at

Pumping in Fossil Plants

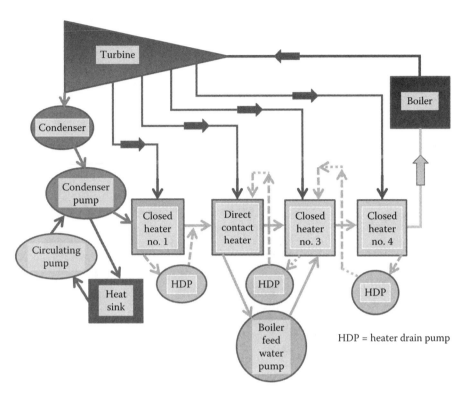

FIGURE 4.5
(See color insert.) Open loop feed water cycle with several heaters.

the cavitation break point, it is not applicable for vertical can-type condensate pumps because of the relatively higher NPSH required, which would provide high enough energy to allow the collapsing vapor bubbles to rapidly erode the first-stage impeller (see Chapter 2, Section 2.3.1).

Condensate from the closed heaters can be pumped to the feed water cycle to avoid both loss of the heat content of this water as well as wasting this water, as shown in Figures 4.4 and 4.5. Piping each heater drain to the next lower pressure heater requires no heater drain pumps, with the lowest pressure heater drained back to the condenser.

Condenser circulating pumps in modern plants are usually of a vertical configuration, of either dry-pit or wet-pit type. The dry-pit type condenser circulating pump operates in a surrounded-by-air environment. The wet-pit type operates either partially or completely surrounded by the pumped water. The choice between these two has been somewhat controversial (Karassik and Carter 1960). Figure 4.6 shows a typical vertical condenser circulating pump.

FIGURE 4.6
Vertical dry-pit mixed-flow condenser circulating pump.

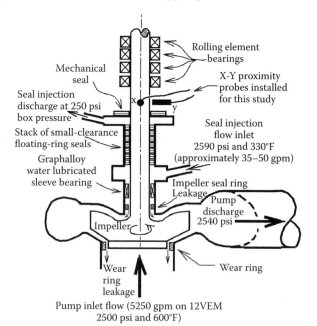

FIGURE 4.7
Cross section of a boiler circulating pump.

Pumping in Fossil Plants

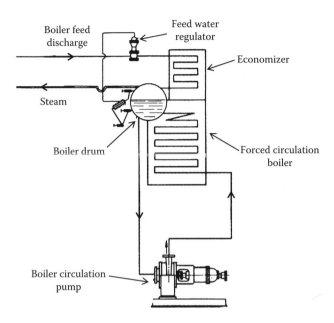

FIGURE 4.8
Boiler forced-circulation cycle.

4.3 Boiler Circulating

An illustration of a *boiler circulating pump* is shown in Figure 4.7. It is from a family of 1950s vintage generating units and is discussed in detail later in Section III as one of the troubleshooting case studies. As boiler sizes increased along with the generating capacity of fossil units, an approach for minimizing the boiler physical dimensions was to employ a number of boiler circulating pumps (typically three or four in parallel) to eliminate total reliance on free convection heat transfer in the boiler. When these pumps are maintained in good running condition, the boiler can operate at full load with one less than the total number of boiler circulating pumps installed in parallel. Advantages touted by proponents of using boiler circulating pumps include (1) smaller diameter boiler tubes, (2) layout freedom of boiler tube arrangements, (3) reduction in number and size of down takes and risers, (4) lower boiler support structure weight, and (5) greater flexibility of operation.

However, the operating conditions for boiler circulating pumps are among the most severe conditions of service in the plant. Figure 4.7 shows the obviously severe operating pressure and temperature combination. Figure 4.8 shows a boiler circulation cycle employing a boiler circulating pump. Karassik and Carter (1960) provide some additional background on these pumps.

5

Pumping in Nuclear Plants

A nuclear reactor converts atomic energy into heat. Figure 5.1 illustrates the *fission* atomic process that is utilized in land- and naval-based nuclear-powered systems. A chain reaction from uranium U-235 pellets in the *fuel rods* is controlled to a steady-state rate of heat production with control rods made of a neutron-absorbing material. *Source rods* containing neutron sources are used to initiate startup of the reactor. Very high velocity neutrons are slowed in the *moderator*, converting their kinetic energy into heat. At full extension the adjustable control rods allow just enough free neutrons for the heat production needed to supply the thermal energy utilized to drive the steam turbine at full load. In addition to the *heated moderator* there must also be a *coolant* to transfer the heat from the moderator to the steam-producing process that powers the steam turbine-generator. The nuclear reactor system is basically equivalent to the boiler in a fossil burning power plant.

In U.S. nuclear power plants the water circulated through the reactor serves both as the moderator and the coolant. Outside the United States the carbon-moderated air-cooled reactor type is also employed (e.g., Chernobyl). In the United States there are two types of water-cooled reactors employed in power plants: the pressurized water reactor (PWR) (Figure 5.2a) and the boiling water reactor (BWR) (Figure 5.2b). PWRs make up about two-thirds and BWRs one-third of U.S. commercial nuclear power. Naval applications all utilize PWRs.

The PWR system has a highly pressurized primary loop to prevent boiling in the reactor, which is hermetically separated from the secondary loop. The reactor-produced heat in the primary loop is transferred to the steam-producing secondary loop through a large closed heat exchanger referred to as the *steam generator*. Not reflected in Figure 5.2a, the primary loop of a commercial PWR has at least single redundancy as illustrated by Makay et al. (1972) (Figure 5.3). When a reactor primary coolant pump (PCP) fails to pump in service for whatever reason, the remaining PCPs must suffice to safely reduce the load as needed while additionally accommodating the backflow through the nonpumping PCP.

Figure 5.4 shows detailed cutaway illustrations of the three main PWR components: reactor, primary coolant pump, and steam generator. A dimensional illustration of a PWR primary coolant pump is shown in Chapter 2, Figure 2.26a. In a PWR system, since the steam piped to the main turbine is not radioactive, the turbine does not need to be housed within the containment vessel, an obvious advantage for construction as well as for

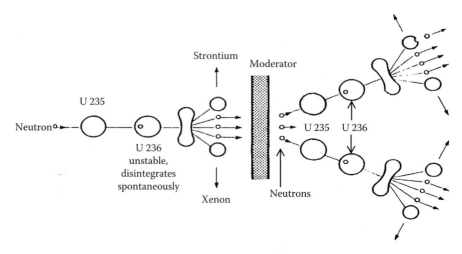

FIGURE 5.1
The nuclear fission process that converts atomic energy into heat.

turbine-generator maintenance. The BWR system is conceptually simpler than the PWR system since it has but one loop with the steam produced within the reactor. The tradeoff is consequently that the BWR system has the turbine housed within the containment vessel.

The steam turbine-generator portion of a nuclear powered unit is not appreciatively different from that of a fossil burning unit. However, in the United States all commercial nuclear unit steam turbines operate at 1800 rpm, that is, 4-pole machines on a 60 Hz grid. Therefore, realizing that Power = Torque × Speed, the U.S. nuclear plant steam turbine-generator physical size is significantly larger than that of a fossil plant 3600 rpm turbine-generator unit of the same power rating. Figure 5.5 shows a 1000 MW 1800 rpm steam turbine-generator for a nuclear plant. In Europe and other 50 Hz grids, some of the nuclear units operate at 1500 rpm (4-pole) and some at 3000 rpm (2-pole). The rotational speed reflects the additional emphasis on safety-accorded nuclear plants. Basically, rotational-speed-dependent rotor stresses increase with the square of the speed and the overall propensity for potential vibration problems increases with speed as well.

The attention to pumps in this chapter focuses primarily on those that are particular to nuclear powered plants and safety related. Figure 5.6 summarizes those major nuclear plant pumps. Given that aging fossil-fired plants are continuously being decommissioned and that sustainable energy options (e.g., wind, solar, micro-hydro) will not nearly fill all energy needs for the foreseeable future, if ever, nuclear power will not be abolished, at least not in the United States. Adams et al. (2004) recommended to the U.S. Nuclear Regulatory Commission future areas of research to further improve nuclear power safety as summarized in Figure 5.7.

Pumping in Nuclear Plants

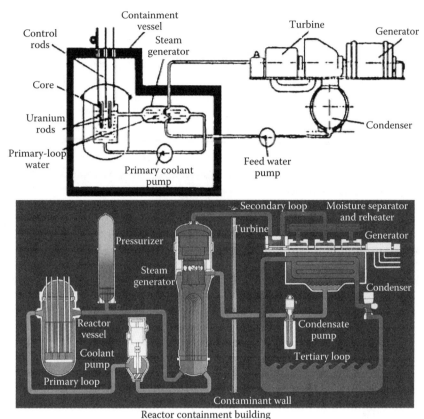

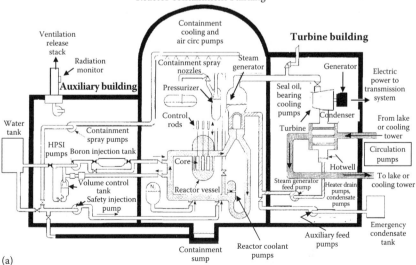

FIGURE 5.2
Essentials of nuclear power plants: (a) PWR system. (*Continued*)

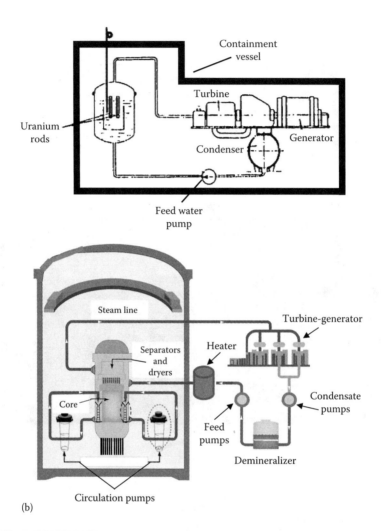

FIGURE 5.2 (CONTINUED)
Essentials of nuclear power plants: (b) BWR system.

5.1 Pressurized Water Reactor Primary Reactor Coolant

The PWR *primary coolant pump* (PCP or RCP) is the most crucial safety-related pump in the plant. PWRs thus typically have two or more coolant loops with at least one PCP operating in each coolant loop as illustrated in Figure 5.3. In the unlikely event that both the auxiliary electric power to the plant and both diesel-driven backup auxiliary power units fail to deliver power to the PCP motors, the quite large flywheel on the top of PCP motors is sized large enough to provide a sufficiently long coast downtime of PCPs

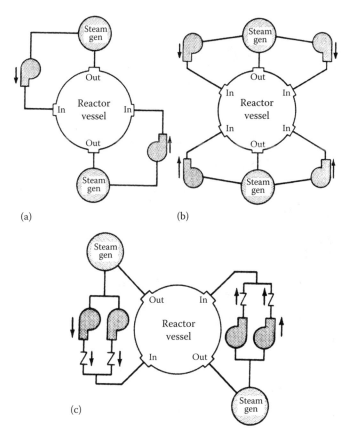

FIGURE 5.3
Schematic diagrams of typical primary coolant loops for PWRs. (a) One pump per loop, (b) two pumps per loop, and (c) two parallel pumps per loop.

to provide sufficient cooling to allow the control rods to be fully inserted into the reactor to safely cease major heat production. The upper motor assembly contains an antireverse rotation piece to prevent the PCP from "turbining" when motor power is interrupted. The PCP shown in Figures 2.26 and 5.6 is a later version of that shown in Figure 5.4b. The difference is in the coupling arrangement between motor and pump. In the later version, an easily removable coupling spool piece has been added to allow inspection and maintenance of the pump shaft seal system without removing the motor, saving considerable maintenance downtime by minimizing the number of operations involved in pump shaft seal inspection.

In both versions, the rigidly coupled rotor is radially supported by three journal bearings, two in the motor and one (primary-loop water-lubricated) near the pump impeller. Thus the journal bearing loads are statically indeterminate. Therefore, the journal bearing loads will not only depend upon pump-hydraulic static radial rotor forces (see Chapter 1, Figures 1.12 and 1.13)

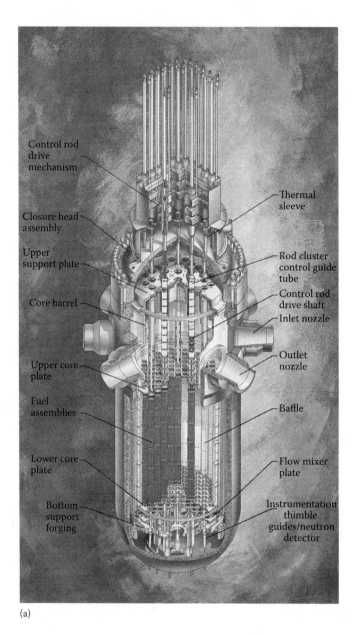

FIGURE 5.4
(See color insert.) PWR (a) reactor. (*Continued*)

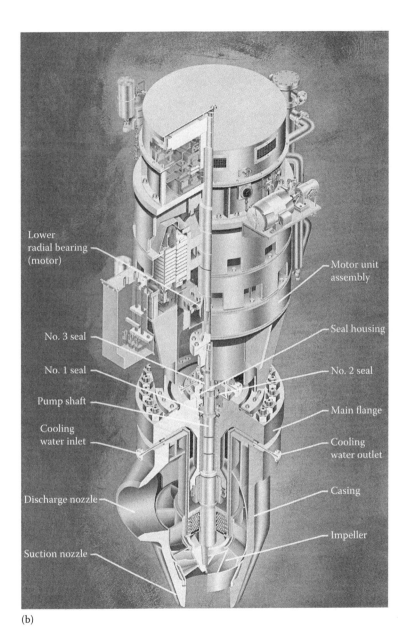

(b)

FIGURE 5.4 (CONTINUED)
(See color insert.) PWR (b) primary coolant pump. (*Continued*)

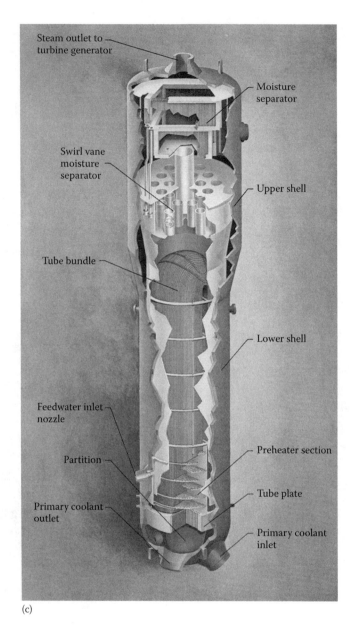

(c)

FIGURE 5.4 (CONTINUED)
(See color insert.) PWR (c) steam generator.

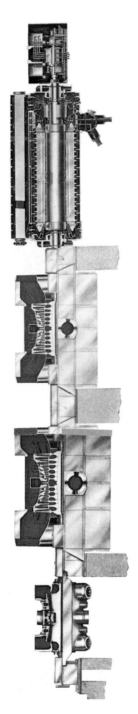

FIGURE 5.5
(See color insert.) 1000 MW 1800 rpm steam turbine-generator for U.S. nuclear plant.

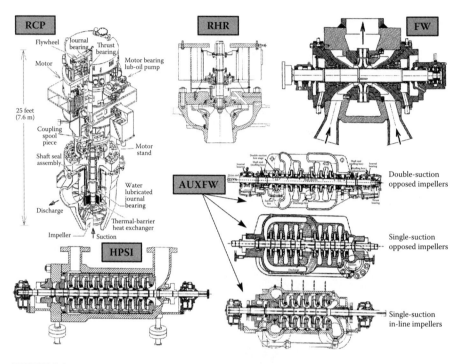

FIGURE 5.6
Nuclear power plant major pumps: Reactor coolant pump (RCP), residual heat removal (RHR), feed water (FW), auxiliary feed water (AUXFW), high pressure safety injection (HPSI).

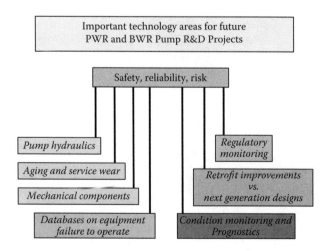

FIGURE 5.7
Recommended research areas to strengthen nuclear plant safety.

and electric motor static radial rotor forces, but also depend upon the radial alignment tolerance between the three journal bearings and rotor flexibility. Combining this fact with the PCP having a vertical rotor, and therefore the total rotor weight being carried by the axial thrust bearing and not the journal bearings, the resultant journal bearing loads are quite random. Adams (2010) clearly explains and demonstrates with troubleshooting case studies that journal bearing rotor dynamic stiffness and damping properties are strong functions of bearing static loads. So rotor vibration characteristics (e.g., critical speeds, instability thresholds) also vary widely over time on the same PCP. Figure 5.8 illustrates the sources of bearing loads on a PCP rotor. Of all the U.S. PCP manufacturers and non-U.S. PCP manufacturers, only one manufacturer (German)

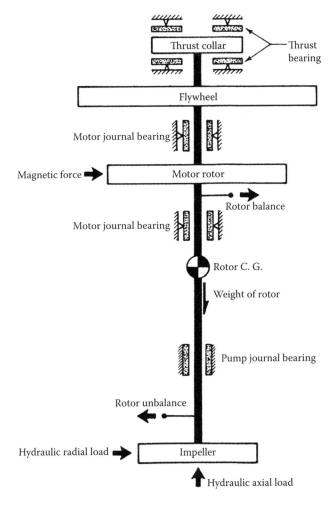

FIGURE 5.8
Sources of bearing loads on a primary coolant pump rotor.

has not used the three-journal-bearing rigidly coupled rotor with statically indeterminate journal bearing loads. That configuration is schematically illustrated in Figure 5.9 (Makay et al. 1972). As shown, the German design has two flexibly coupled rotors. Thus all bearing loads are statically determinant. So its sensitivity/variability over time, of journal bearing loads and rotor dynamic characteristics to manufacturing and assembly tolerances, is much less than for the other manufacturers' three-journal-bearing rigidly coupled rotors.

Since nearly all the PCPs have now been in service for many decades, aging becomes a significant matter of concern. Much has been learned by technologists on critical design aspects of these PCPs since their commissioning, particularly in (a) journal bearing load determination, (b) rotor dynamic characteristics, (c) unbalance rotor vibration, (d) seal failures, (e) shaft failures, (f) seismic robustness, and (g) monitoring and diagnostics (Makay and Adams 1979). As power producers apply for operating license extensions for aging nuclear power plants, the critical technologies such as illustrated in Figure 5.7 for nuclear safety-related pumps, need to be comprehensively utilized in regulatory decisions for operating license extensions. At the same time, it is quite relevant to keep in mind that the original developers/designers/evaluators of these PCPs have either passed away or are very long into retirement, the author not excluded.

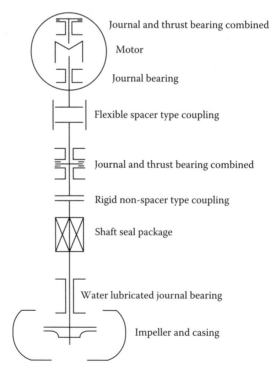

FIGURE 5.9
Schematic of German manufacturer's PCP.

5.2 Feed Water and Auxiliary Feed Water

Figure 1.7 (Chapter 1) shows the cutaway of a typical single-stage double-suction nuclear feed water pump with a rotational speed around 5000 rpm. Typically two such 50% feed pumps are installed and turbine driven. An additional motor driven 50% feed water pump may also be used for startup and as a reserve or standby pump, although a small startup feed water pump is generally employed for startup and shutdown. Less common configurations used are more similar to boiler feed pumps, which can be either turbine or motor driven. The feed water and auxiliary feed water pumps of both PWR and BWR systems are among the most critical safety-related pumps in a nuclear power plant. In the PWR system, the feed water/steam system is within the secondary loop and thus is hermetically separated from the primary loop water radiation, except when significant steam generator tube leaks occur, which will be detected by radiation sensors in the steam turbine. As Figure 5.6 illustrates, *auxiliary feed pumps* are more akin to typical multi-stage boiler feed pumps of fossil plants.

5.3 Residual Decay Heat Removal

As Dahlheimer et al. (1984) explain, the primary function of the residual heat removal system (RHRS) is to transfer heat energy from the reactor core and reactor coolant system (RCS) during plant cooldown and refueling operations. The RCS temperature is reduced to 140°F (60°C) within 20 hours following reactor shutdown. The RHRS may also be used to transfer refueling water between the refueling cavity and the refueling water storage tank at the beginning and end of refueling operations. The residual heat removal pumps are also utilized as part of the *safety injection system* for emergency core cooling in the event of a loss-of-coolant accident (LOCA).

5.4 High Pressure Safety Injection and Charging

Figure 5.6 shows the cutaway of a *high pressure safety injection* (HPSI) pump. The safety injection system (SIS) has many purposes, the most important being to provide emergency reactor core cooling in case of a loss-of-coolant accident (LOCA). The SIS also adds negative reactivity with injection of borated water to meet shutdown requirements and/or to compensate for the reactivity increase caused by cooldown transients such as from a steam

FIGURE 5.10
Charging pump.

line break. The SIS positive displacement hydro-test pumps also provide a backup source of reactor coolant pump seal injection water. In the unlikely event of a LOCA, the SIS is designed to limit increases in fuel clad temperatures, core geometry distortion, and metal–water reaction for all break sizes. For the more probable break sizes less than or equal to 5 inch (12.7 cm) inside diameter, the SIS is designed to minimize core damage by providing flow to the core that is sufficient to prevent the mass depletion-related uncovering of the core. The system is designed to provide not only emergency core cooling but also continued cooling during the long-term phase following the accident. High pressure safety injection water is provided by separate high-head pumps while lower pressure injection water is supplied by the residual heat removal (RHR) pumps. Passive accumulator tanks are located inside the containment to provide for fast injection of water following a LOCA.

Charging pumps circulate reactor primary coolant, at the pumping maximum temperature of 284°C, to the treatment plant. These pumps also limit temperatures in a BWR pressure vessel by forced circulation within a closed loop. The charging pump configuration is a motor driven, flexibly coupled, horizontal centerline, radially/vertically split volute casing centrifugal pump with a double-suction first-stage impeller (Figure 5.10).

5.5 Boiling Water Reactor Main Circulating

A BWR recirculation system is used for (a) control of reactor power level through the variable speed recirculation pumps, (b) cooling of the reactor

Pumping in Nuclear Plants 109

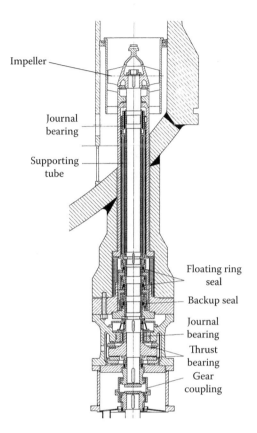

FIGURE 5.11
BWR main circulating pump.

during part load operation, and (c) emergency makeup and chemical purification. The recirculation system consists of two recirculation loops, each consisting of piping from the reactor to a recirculation pump and return line back to the reactor. The recirculation pumps are driven by *variable speed motors*. As the speed increases, more voids are swept from the core, resulting in more thermal neutrons being produced. This increases power produced. The pumps are installed in the annular region between the core barrel and the outside reactor vessel as shown in Figure 5.11.

5.6 Quarterly Testing of Standby Safety Pumps

As exposed by Casada and Adams (1991) and Adams (1992), the quarterly short-duration testing of standby safety pumps can quite possibly produce more wear and tear than if these pumps were operated all of the time continuously at their design-flow point. These pumps include (a) auxiliary feed water, (b) high pressure safety injection, and (c) decay heat pumps (see Figure 5.6). The fundamental stressor is the typical quarterly testing of these pumps at flows near shut-off condition, that is, at flows that are considerably below their best-efficiency-flow design point. Chapter 3, Section 3.1.3 highlights the torturing effects at low-flow operation. But plant designers apparently did not seriously think about this due to the relative short duration of quarterly operability tests. And, of course, if a safety-related backup pump is in fact operated in a real emergency like a LOCA condition, it will be operating at its full flow operating condition. So quarterly testing at near shut-off flow does not really validate or confirm pump performance for a true emergency condition.

Section III

Troubleshooting Case Studies

6

Boiler Feed Pump Rotor Unbalance and Critical Speeds

The most common cause of excessive rotor vibration is mass unbalance in the rotor, and the primary symptom is, of course, excessive once-per-rev (synchronous) vibration. Vibration excited by residual rotor unbalance is always present in all rotors at all operating speeds, because it is impossible to make any rotor perfectly mass balanced. Therefore, the objective concerning rotor unbalance excited vibration is limiting it to allowable levels, not the impossibility of total elimination. Figure 3.25 and Table 3.1 (see Chapter 3) provide guidelines categorizing whether residual vibration amplitudes are within acceptable limits. When vibration levels are deemed excessive and it has been established that the excitation is from rotor unbalance, the proper corrective course of action is often simply to rebalance the rotor. However, in centrifugal pumps, strong synchronous vibration can also originate from sources other than rotor mass unbalance, most notably centrifugal pump hydraulic forces (see Section 3.2). Therefore, it is readily apparent just from the symptoms associated with excessive rotor unbalance that identification of specific root causes of excessive pump vibration remains an inexact science.

The most frequent rotor-balancing job is the in-service quick balance correction. A balance correction weight is placed on the rotor at a readily accessible location. The objective of such *single-plane in-service balancing shots* is to reduce the maximum vibration levels. It is not intended nor is it feasible that such a single-plane balance shot provide the high quality degree of rotor balance that is achievable when the removed bare rotor is factory component balanced, impeller-by-impeller, and then balanced fully assembled in a precision balancing machine.

The root cause for excessive synchronous vibration can also be other than the rotor being too far out of balance or the hydraulic result of poor impeller cast vane accuracy. If the operating speed is too close to an inadequately damped resonance condition (i.e., critical speed; see Chapter 3, Figure 3.7), the synchronous vibration level can be excessive. The case studies presented in this chapter are not of the category where routine in-service rebalancing of the rotor is the solution to the excessive vibration problem. Each case study presented here typifies the more difficult ones to solve where routine rebalancing does not solve the problem. As these cases demonstrate, identification of both root cause(s) and the most cost-effective solution(s) or fixes can be enormously aided by using analysis models. Adams (2010) provides detailed treatments on rotor vibration modeling for troubleshooting.

6.1 Case 1

The rotor sectional view shown in Figure 6.1 is from a four-stage boiler feed pump (BFP). In the power plant of this case study, the BFPs are installed as variable speed units with operating speeds from 3000 to 6000 rpm, each with an induction motor drive through a variable speed fluid coupling. In this plant, the BFPs are all "50%" pumps, so when the main stream turbo-generator is at 100% full load power output, two such pumps are operating at their full design H-Q operating condition. The plant in this case houses four 500 MW generating units, each having three 50% BFPs installed (i.e., one extra 50% BFP on standby), for a total of 12 boiler feed pumps of the same configuration. Full-load operating ranges for each 50% BFP is 5250 to 5975 rpm, 684 to 1035 m^3/hr, and differential pressures from 13.4 to 21.0 MPa.

The BFPs at this plant had experienced a long history of failures, with typical operating times between overhauls under 10,000 hours, with the attendant excessive monetary costs. Based on the operating experience at other power plants employing the same BFP configuration with quite similar operating ranges, these BFPs should have been running satisfactorily for over 40,000 hours between overhauls. Using vibration velocity peak monitored at the outboard bearing bracket, these BFPs were usually taken out of service for overhaul when vibration levels exceeded 15 mm/sec (0.6 in/sec). To wait longer significantly increased the overhaul rebuild cost, that is, more damage. The dominant vibration frequency was synchronous.

The author's preliminary diagnosis was that these pumps were operating quite near a critical speed and that the resonance vibration resulting from this worked to accelerate the wearing open of interstage sealing ring radial clearances. As these interstage clearances wear open, the overall vibration damping capacity diminishes significantly, typically leading to a continuous growth of vibration levels. To confirm this preliminary diagnosis, the

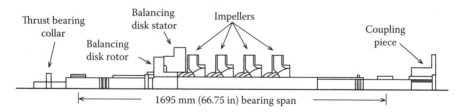

FIGURE 6.1
Rotor sectional view for a four-stage boiler feed water pump.

author developed a finite-element-based rotor-vibration computer model for this BFP configuration to compute lateral rotor vibration unbalance response versus rpm, as exampled in Figure 3.7. The manufacturer of the pump provided a nominally dimensioned layout of the assembled pump, including weight and inertia for concentrated masses (impellers, balancing disk, thrust bearing collar, coupling piece, and shaft sleeves). The pump OEM also provided detailed geometric dimensions for the journal bearings, interstage radial seals, and other close-clearance radial annular gaps. This cooperation by the pump OEM greatly expedited the development of the rotor-vibration computer model, eliminating the need to take extensive dimension measurements from one of the BFPs at the plant or repair shop in Australia.

The radial annular gaps have clearance dimensions that are quite small and are formed by the small difference between a bore inside diameter (ID) and an outside diameter (OD), each with tolerances. The size of each of these small radial clearance gaps is very influential on the respective bearing or seal stiffness, damping, and inertia coefficients, and thus very influential on the computed results for rotor vibration response. However, these small radial gaps vary percentage-wise significantly and randomly because of their respective ID and OD manufacturing tolerances plus any wearing open due to in-service use. BFPs are thus one of the most challenging rotating machinery types to accurately model and analyze for rotor vibration. The net result is that even in the easiest of cases, a realistic rotor vibration analysis for troubleshooting purposes (as opposed to design purposes) requires several trial input cases to get the model predictions to reasonably portray the vibration problem the machine is exhibiting. By iterating the model inputs per radial-clearance manufacturing tolerances and allowances for wear, a set of inputs is sought that produce rotor vibration response predictions that concur with the machine's vibration behavior. When or if a good agreement model is so obtained, it is referred to as the *calibrated model*. Through such computer simulations, the calibrated model can then be used to explore the relative benefits of various fixes or retrofit scenarios.

A calibrated model was not initially achieved for this pump vibration problem in that all reasonable model variations for input dimensions failed to produce predicted unbalance responses having a resonance peak below 8000 rpm, which is considerably above the maximum operating speed. Since the power plant in this case was a considerable distance outside the United States, a visit to the plant had not initially been planned. However, given the failure of all initial computer model variations to replicate or explain the BFP excessive vibration problem, a trip to the plant south of Melbourne was undertaken to study the pumps firsthand.

Poor hydraulic conditions in BFPs, such as from inaccurate impeller castings, can produce strong synchronous rotor vibrations, so several of the impellers were inspected for such casting inaccuracies. In the course of further searching for the vibration problem root cause, a number of serious deficiencies were uncovered in the local BFP overhaul and repair shop's methods and procedures, all of which collectively might have accounted for the vibration problem. Luckily, on the last day of the planned one-week visit to the plant, the root cause was discovered, but it could have been easily overlooked. In the process of discussing installation-after-overhaul details with a mechanic at the plant who was reinstalling a just-repaired BFP, it was revealed to the author that between the inner journal bearing half shells and the axially split outer bearing housings there was a clearance of about 0.001 in (0.025 mm) into which a gasket was interposed and compressed as the two housing halves were tightly bolted together. This use of gaskets had been discontinued many years earlier in most U.S. power plants. The net result of the interposed gasket was to reduce the effective bearing stiffness to a value significantly below the range that had been reasonably assumed in the initial (unsuccessful) attempts to develop a calibrated rotor vibration computer model. When the gasket-clearance effect was incorporated into the computer model inputs, a resonance critical speed peak showed up right within the normal operating speed range.

An analysis study was made to compute critical speed as a function bearing stiffness, using a stiffness value range consistent with the interposed gasket. A summary of the results for this analysis is shown in Figure 6.2. A bearing stiffness value of 100,000 lb/in placed the critical speed right at the normal full load operating speed range. The variability of gasket compressive stiffness also explained the plant's experience with the excessive vibration fading "in and out" over time.

The gasket stiffness is in series with the bearing oil film's in-parallel stiffness and damping characteristics. Since the gasket stiffness is much less than the journal bearing oil-film stiffness, the gasket also reduces considerably the damping action in the oil films, further acerbating the vibration problem. The use of a gasket between the bearing inner shell and outer housing was clearly the smoking gun, placing the critical speed near the normal full-load operating speed while depriving the attendant resonance of reasonable bearing damping. The bearings were reinstalled with metal shims to provide a bearing pinch of about 1 mil (one thousandths of an inch).

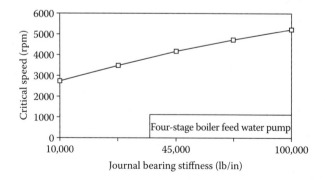

FIGURE 6.2
Critical speed versus bearing stiffness including the interposed gasket.

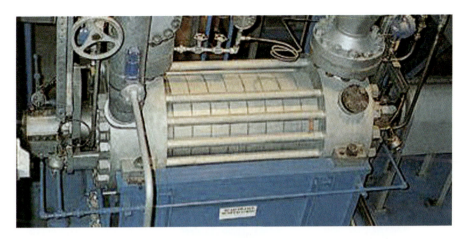

FIGURE 1.16
Radially split ring-section feed water pump.

FIGURE 2.11
Pump shaft after sudden seizure of lubrication-starved gear coupling.

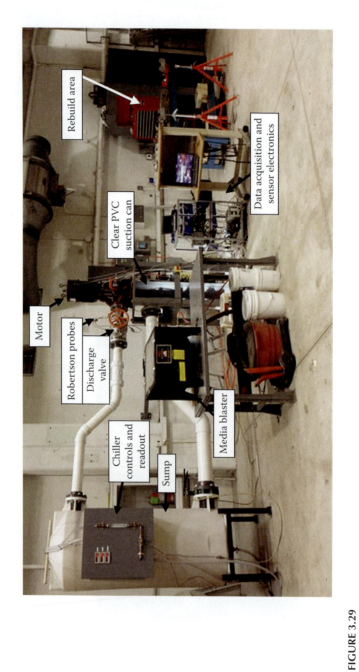

FIGURE 3.29
CWRU multistage centrifugal pump research test loop.

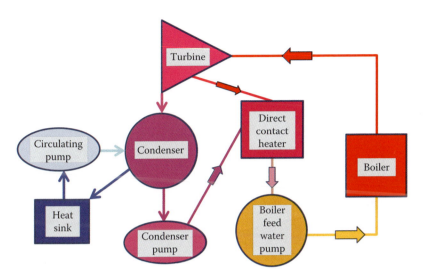

FIGURE 4.1
Elementary steam power unit cycle.

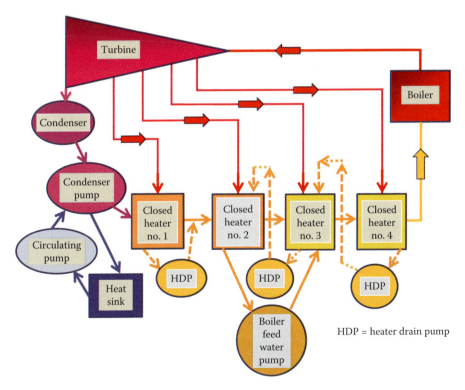

HDP = heater drain pump

FIGURE 4.4
Closed loop feed water cycle with several heaters.

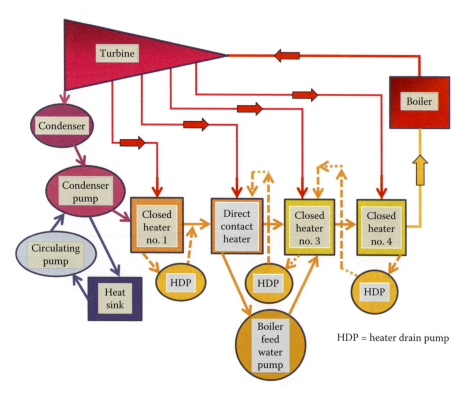

FIGURE 4.5
Open loop feed water cycle with several heaters.

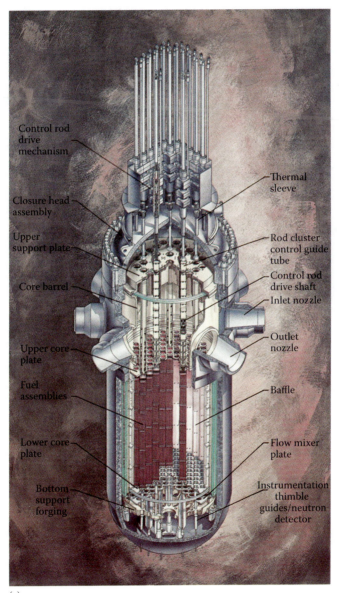

(a)

FIGURE 5.4
PWR (a) reactor. (*Continued*)

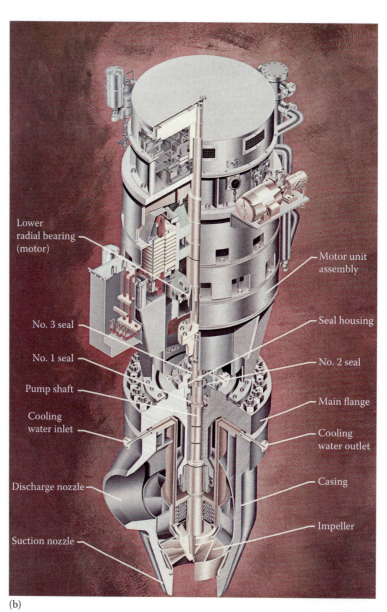

(b)

FIGURE 5.4 (CONTINUED)
PWR (b) primary coolant pump.   (*Continued*)

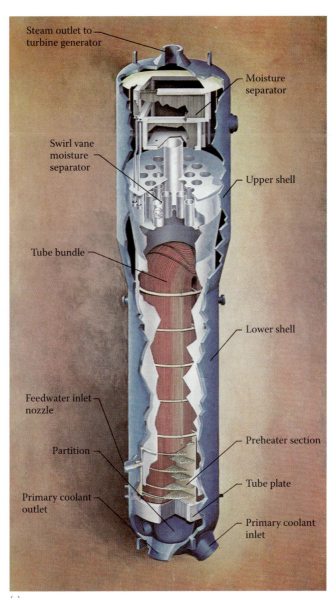

(c)

FIGURE 5.4 (CONTINUED)
PWR (c) steam generator.

FIGURE 5.5
1000 MW 1800 rpm steam turbine-generator for U.S. nuclear plant.

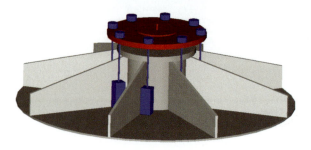

FIGURE 9.4
Vibration absorber (red and blue) atop pump motor.

Boiler Feed Pump Rotor Unbalance and Critical Speeds

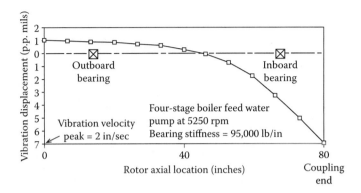

FIGURE 6.3
Critical speed rotor response shape with typical unbalances.

A simplified planar view of the computed BFP nonplanar critical-speed rotor-response shape is shown in Figure 6.3, which graphically "flattens" the response shape into a plane. This is helpful in showing the rotor axial locations where residual rotor mass unbalance will have the most effect in exciting the critical speed resonance vibration. The obvious conclusion drawn from the computed unbalance response shape shown in Figure 6.3 is that coupling unbalance probably contributed significantly to this vibration problem, because the repair shop's rotor balancing procedure, as witnessed, was inadequate in several areas, particularly in their handling of the coupling. The flexible couplings employed on these BFPs are of the diaphragm type and are well suited to the BFP application, being more reliable than gear couplings that require maintaining lubrication (see Chapter 2, Section 2.4.2). With a properly functioning flexible coupling, the BFP is sufficiently isolated from the driver (lateral vibration-wise) so that the rotor vibration analysis model in this was justifiably terminated at the pump half of the coupling. Experience has shown this to be well justified.

6.2 Case 2

A second BFP vibration case study presented here involves the BFP shown assembled in Figure 6.4. It is similar in size and capacity to that in the Case 1 study (Figure 6.1), being a 50% pump for a 430 MW steam turbo-generator unit. The BFP shown in Figure 6.4 is actually a three-stage pump for boiler feed, but has a small fourth stage (called a "kicker stage") that is to supply high-pressure injection water at pressures above feed water pressure.

This BFP was observed to have a critical speed at 5150 rpm, although the manufacturer's design analyses did not support this observation. This is a

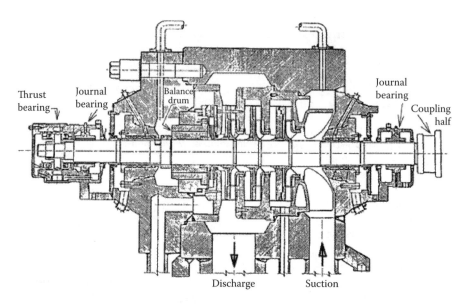

FIGURE 6.4
Assembly view of a boiler feed water pump.

variable speed pump with a maximum operating speed of 6000 rpm. The 5150 rpm critical speed was in the frequently used operating speed range and produced excessive vibration levels, primarily at the inboard end of the rotor, that is, coupling end at suction inlet. A clue was supplied by Dr. Elemer Makay (see "In Memoriam"). At a number of power plants employing the same BFP design, he observed BFP inboard journal bearing distress in the top half of the bearing bore. This bearing distress was consistently centered about 10° rotation direction from the top center. The journal bearings were of a design employing a relieved top-half pocket. The specific elbow geometry of the pump inlet piping suggested to Makay that there was consequently a significant upward hydraulic static impeller force on the suction end (inboard end) of the rotor.

A finite-element-based rotor unbalance vibration response model was developed by the author from detailed OEM information supplied by the electric power company owner of the plant. A lengthy double-nested iteration study was undertaken in which an upward static rotor force was applied on the rotor model at the suction-stage impeller. Through a trial-and-error iteration, this upward static radial force was directed so as to produce an inboard journal eccentricity direction of 10° rotation from the journal bearing top center, motivated by the bearing distress observations of Makay. From each of several values assumed for this force, a set of journal bearing static loads were calculated. A set of bearing stiffness and damping coefficients were in turn calculated for each set of bearing static loads. Each set of bearing stiffness and damping coefficients were then used as inputs to the finite-element based unbalance response model to compute rotor vibration response versus revolutions per minute using a typical set of rotor unbalances.

Through several iterations, a force of 3477 pounds yielded journal bearing rotor dynamic coefficients that predicted an unbalance-excited critical speed of 5150 rpm. Furthermore, at this predicted 5150 rpm critical speed, the rotor vibration response shape showed high inboard (coupling end) vibration levels as observed on the BFPs at the plant. In fact, the critical speed rotor vibration response shape was very similar to that shown in Figure 6.3 for Case 1. The problem was eliminated by retrofitting a different journal bearing configuration that shifted the critical speed considerably above the 6000 rpm maximum operating speed.

6.3 Case 3

The boiler feed pump shown in Figure 6.5 experienced a number of forced outages that were accompanied by excessive vibration levels. One of these outages involved a complete through-fracture of the pump shaft just adjacent to the balancing drum runner. The author was retained to diagnose the root cause(s) and develop a cost-effective fix. This pump was not equipped with shaft targeting noncontacting displacement proximity probes. So the author's first step was to retrofit two proximity probes 90° apart near each pump journal bearing to obtain shaft vibration displacement measurements adequate for successful root cause diagnoses. These four retrofitted proximity probes were installed in parallel with four velocity pickups to capture any proximity probe mounting motions. Figure 6.6 shows the outboard end of the pump with the author's retrofitted vibration X and Y proximity probes and velocity pickups.

A parallel task was undertaken to develop a rotor unbalance vibration response computer model for this pump. Computer model prediction results are shown in Figure 6.7, and predict a critical speed at 5250 rpm, right near the normal full load operating speed. Subsequent to these model predictions, the unit was restarted and all eight channels of newly installed vibration channels were recorded as a function of pump rotational speed during roll-up. A sample of these vibration measurements is plotted in Figure 6.8. These rotor vibration measurements clearly show a vibration peak at about 5100 rpm, quite close to the premeasurement predicted 5250 rpm critical speed. This critical speed was judged a strong contributing factor to the excessive pump vibrations and associated outages. The author engineered wear-ring surface geometry modifications for this pump (Lomakin shallow grooves; see Chapter 2, Figure 2.23b) to shift the critical speed well above the operating speed range.

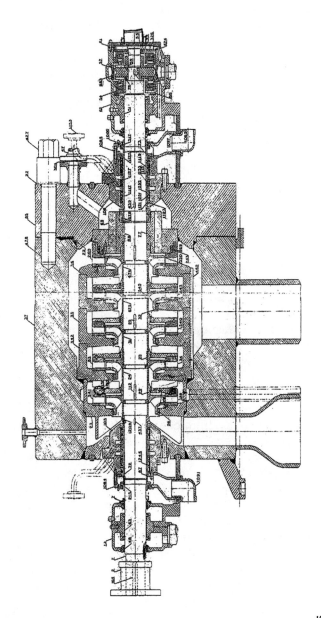

FIGURE 6.5
Variable-speed boiler feed pump of a 600 MW generating unit.

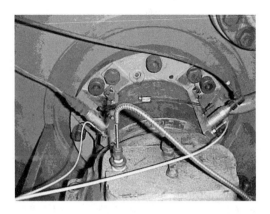

FIGURE 6.6
Retrofitted proximity probes and velocity pickups.

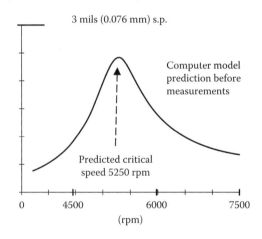

FIGURE 6.7
Computer model prediction before measurements.

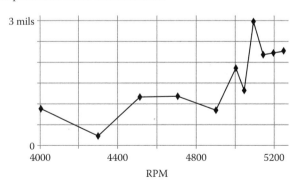

FIGURE 6.8
Boiler feed pump shaft vibration measurements.

6.4 Summary

The three boiler feed pump cases presented in this chapter demonstrate the considerable challenges in developing good predictive rotor vibration models for multistage centrifugal pumps. These challenges arise from two sources. The first is the multiplicity of liquid-filled annular rotor-stator small-clearance radial gaps that dominate the vibration characteristics of such machines, combined with the dimensional variability of these small radial gaps from ID and OD manufacturing tolerances and in-service wear. Second, the potentially large and uncertain hydraulic radial static impeller forces, which vary with a pump's operating speed and points over its head–capacity curve (see Chapter 1, Equation 1.10), introduce considerable uncertainty in radial bearing static loads. Since a journal bearing's rotor dynamic characteristics are strong functions of its static load, the inherent uncertainty of impeller static radial forces adds to the uncertainty for rotor vibration modeling and problem diagnoses. These case studies demonstrate the diligent persistence required to isolate the root cause(s) in cases where simply rebalancing the rotor does not solve the problem.

7

Nuclear Feed Pump Cyclic Thermal Rotor Bow

7.1 Background on Cyclic Vibration Symptom

This troubleshooting case was presented for the author's investigation after the plant owner company conducted an extensive but unsuccessful in-company project to alleviate a vibration problem in all the feed water pumps in a plant housing two 1150 MW pressurized water reactor (PWR) generating units. Each PWR unit has two 50% feed pumps with a spare feed pump on site. A cross-sectional layout of the feed pump configuration is shown in Figure 7.1. All four operating feed pumps had experienced cyclic rotor vibration spikes that were synchronized with the seal injection water flow control.

Only after several months of unsuccessful in-company vibration measurements and troubleshooting diagnoses of this excessive vibration problem, the author was retained by the power company owner to see if a rotor vibration computer model analysis could identify the root cause of the excessive vibration. A more detailed documentation of this troubleshooting case is reported by Adams and Gates (2002) where the plant and pump OEM are identified. The correlation between pump seal injection water control and vibration signals is shown in the 50-minute sample of vibration data in Figure 7.2.

7.2 Rotor Vibration Analyses

Exhaustive computer rotor vibration analyses of this pump were conducted by the author in a manner similar to the successful use of computer modelling to assist in the three boiler feed pump troubleshooting cases presented in Chapter 6. The rotor vibration computer analyses included investigations for root causes from critical speed resonances and instability self-excited rotor vibration phenomena. It was no surprise to the author that these analyses eliminated critical speeds and self-excited vibration phenomena as likely root causes, but the plant insisted upon these analyses as the first step.

124 Power Plant Centrifugal Pumps

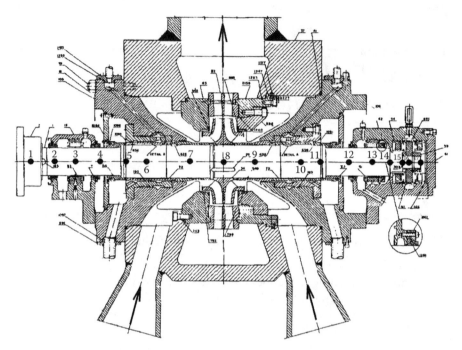

FIGURE 7.1
Nuclear feed water pump; analysis mass stations are numbered.

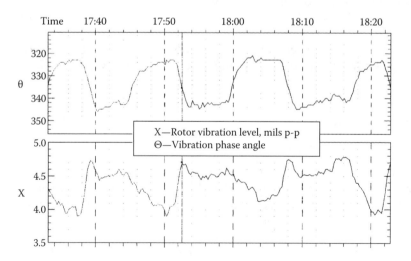

FIGURE 7.2
Fifty-minute vibration record from feed water B Pump Unit 2.

7.3 Cyclic Thermal Bow Analysis

With those analyses out of the way, the author was free to study if a cyclic thermal bowing of the rotor could be the root cause of the problem that could explain the plant's pump rotor vibration. The 10°F cyclic seal injection differential temperature swings synchronized with the 15 min/cycle rotor vibration characteristic shown in Figure 7.2. A close examination of this pump configuration in Figure 7.1 shows a typical arrangement employing shaft sleeves to form the rotating parts of the shaft seals. There are two mating sleeves on each axial side of the double-suction impeller. Each of these two-sleeve combinations was modeled by a single hollow cylinder of nominal length. A calculated 10°F differential thermal expansion for the two-sleeve model (in steel) was computed to be 1.2 mils (0.03 mm), which was calculated to impose a 23,000 pound (10,250 N) compressive force on the sleeves, since the much higher cross-sectional area of the shaft virtually prevents this differential thermal expansion. Under perfect manufacturing and assembly conditions (i.e., no tolerances), the compressive restraining force would be co-axial with the shaft centerline (i.e., best-case scenario). Under a worst-case scenario (possible), the axial restraining force would be centered at the outer radius of the cylinder ($R \cong 3.5$ in, 89 mm). For a representative bending moment calculation, the intermediate value of $R/2$ was used. Shaft compressive force was accordingly calculated to yield a shaft bending moment of 40,250 in-lb (1.75 × 23,000), 4554 N-m.

7.4 Shop Cyclic Thermal Test and Low-Cost Fix

As illustrated in Figure 7.3, the bending moment was calculated to cause a 3.8 mil (0.097 mm) transient thermal bow of the shaft. This result was initially not believed by the client. So as a prudent next step, the client shop tested the plant's spare feed pump rotor on the rotor balancing machine at the pump OEM's repair shop with specially installed locally placed heaters on the shaft sleeves. This test confirmed the author's analysis results shown in Figure 7.3.

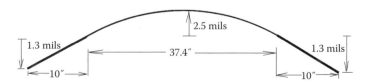

FIGURE 7.3
Computed shaft thermal bow by sleeve-to-shaft differential expansion.

That led to a shaft-sleeve retaining nut modification retrofit, by interposing a compressible annular gasket under each shaft-sleeve retaining nut at both ends of the pump shaft. This low-cost retrofit more evenly distributes the compressive force circumferentially and freely allows the inherent cyclic differential thermal expansion while maintaining the nominal sleeve assembly compressive force. This gasket fix can be theoretically idealized as a "soft spring" with a "large" preload compressive deflection. This retrofit is now installed on all four of the plant's 50% feed water pumps, with total success in eliminating the vibration problem's root cause, as reported by Adams and Gates (2002).

8

Boiler Circulation Pump

8.1 Problem Background

An alternative for fossil-fired boilers for steam power plants was to incorporate boiler circulation pumps (BCPs). This measure significantly reduced the size of the boiler compared to free-convection boiler designs for the same capacity and thus also significantly reduced the boiler first cost. Such BCPs came in a range of sizes matched to the output capacity of the boiler. The pump discussed here was for older generating units, three rated at 130 MW and one rated at 240 MW. Each 130 MW boiler was supplied with three BCPs and the 240 MW boiler with four BCPs. All these units were driven by four-pole induction motors (1750 rpm). All four generating units were designed to operate at full capacity with one BCP not operating, allowing maintenance and rebuilds to take place without backing off on unit generating power output. Figure 8.1 shows a cross section of the BCP for the 130 MW units. All these pumps have a vertical centerline of rotation as illustrated.

This line of BCPs is from a single OEM and has been utilized in several other plants of various sizes, BCP size matched to the specific boiler capacities. The author was retained to do a thorough top-to-bottom investigation of these pumps, because they all had for many years required frequent rebuilds necessitated by the floating ring seal clearance wearing open enough for seal leakage to exceed the capacity of the feed water system to supply seal injection water from a point upstream of the feed water manifold, that is, at a pressure sufficiently higher than the boiler feed water pressure. See Figure 8.1b. BCPs removed from service were shipped to the plant owner's in-house service shop for the overhaul rebuild. There the pump is completely disassembled and not only is the floating-ring seal assembly replaced, but so are other marginally worn internal components, for example, the Graphalloy sleeve bearing. The in-house cost of the rebuild averaged $23,500. By comparison, the OEM's charge would have been approximately $50,000 for a BCP overhaul rebuild. The BCP rebuilds translated into an annual cost of approximately $170,500/year for the plant.

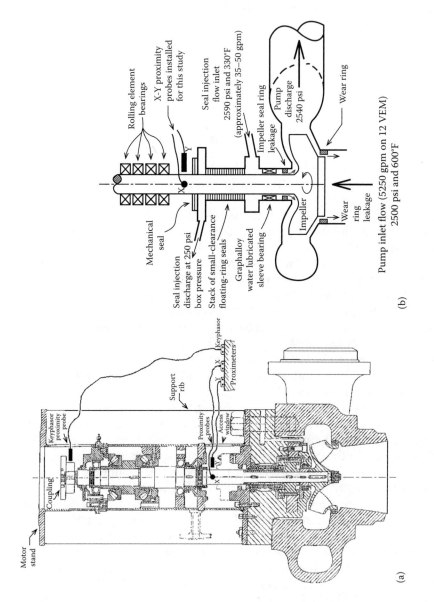

FIGURE 8.1
Boiler circulation pump: (a) cross section and (b) illustrated components.

8.2 Investigation

To gain potentially valuable diagnostic information, special fixtures were devised to measure shaft vibration using displacement proximity probes installed on one of the operating pumps (Figure 8.1). The only readily accessible location for vibration probes was the coupling. Figure 8.1a shows installation of X and Y displacement proximity probes at the only readily shaft-accessible location. It was anticipated that excessive vibration levels might not be the root cause of the fast-wearing seal leakage problem, but that the shaft vibration spectrum might provide revealing clues. A spectrum of one of these shaft vibration displacement signals is plotted in Figure 8.2. Since the rotational centerline is vertical, the appearance of a modest level of N/2 vibration component (1/2 rpm frequency) was not surprising. But what was surprising was the high level of the 6N impeller vane-passing shaft vibration level. This finding led to identification of the root cause of the consistently high wear rate of the floating-ring seal clearance.

Since it is easier to seal liquid water than steam, special measures are employed to ensure that flashing to steam does not occur in the seal leakage flow path. Accordingly, it is necessary to tap seal injection water off the boiler feed pump discharge manifold. Boiler feed pump discharge water is the only water available with a high enough pressure to force flow into the boiler, through the BCP impeller seal ring clearance (see Figure 8.1). Boiler feed pump discharge water is also at a relatively "cool" 330°F and thereby

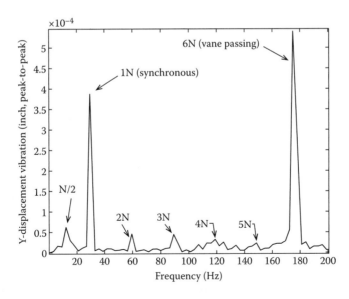

FIGURE 8.2
BCP shaft coupling vibration spectrum.

prevents flashing in the shaft seal leakage flow path. Under proper operation including startup, the seal injection water should be controlled at a nominal 50 psid above boiler pressure to prevent hard-particle laden boiler water from invading the primary shaft seals.

8.3 Floating-Ring Seal Leakage

As a reference point, seal leakage calculations were made by assuming "smooth" sealing annulus surfaces and that seal annulus entrance and exit pressure drops are small compared to the pressure drop within the seal fluid annulus. The basic seal ring annulus configuration is illustrated in Figure 8.3 along with seal leakage calculation formulas.

Five floating rings are contained in the BCP seal assembly, so if each one has an equal pressure drop, the Δp for each ring is (2590 psi − 250 psi)/5 = 468 psid. Using the calculation steps given in Figure 8.3, the seal leakage over the as-new clearance range and for twice the maximum as-new clearance are calculated as follows:

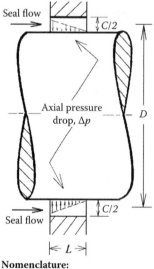

Calculation for turbulent seal flow
1. Estimate U, average axial velocity
2. Compute Reynolds number
   $Re = UD_h/\nu$ and friction factor
   $$f = \frac{0.316}{Re^{0.25}} \text{ (for smooth surfaces)}$$
3. Calculate $U = \left(\dfrac{2\Delta p D_h}{f \rho L}\right)^{1/2}$
4. Iterate, returning to step 2 until desired convergence is achieved.
5. Seal leakage, $Q = UA$

Water at 330°F, 2500 psi:
$\mu = 2.44 \times 10^{-8}$ lbf sec/in^2
$\rho = 7.86 \times 10^{-5}$ lbf sec^2/in^4

Nomenclature:
Δp = pressure drop across sealing fluid annulus, D = diameter of fluid annulus, C = clearance of fluid annulus, L = axial length of fluid annulus, μ = viscosity, ρ = mass density, $\nu = \mu/\rho$, D_h = 4*flow area/wetted perimeter, hydraulic diameter

FIGURE 8.3
Annular sealing gap configuration and leakage calculation.

Minimum as-new clearance, $C = 0.005''$ gives $Q = 11.3$ gpm

Maximum as-new clearance, $C = 0.007''$ gives $Q = 20.3$ gpm

Twice as-new max clearance, $C = 0.014''$ gives $Q = 65.0$ gpm

The seal leakage here increases approximately as the clearance squared.

8.4 Wear Ring Leakage

When one of these pumps is removed to the rebuild shop, the boiler remains in operation. During the time the pump is away for rebuild, the only thing that separates the inside 600°F 2500 psi (316°C 170 bars) boiler water from the surrounding work area is a single isolation valve, hopefully with a good seat! So naturally no one in their right mind is going to stick his or her head down into the impeller vacated area to take an inside micrometer reading of the impeller-eye wear-ring inside diameter. When the simultaneous opportunity of a unit shut down and a BCP removal fortuitously coincided, the author asked for an impeller-eye wear-ring ID measurement. The measurement showed that the impeller wear ring radial clearance was several times the maximum as-new value, worn open from years of use and unattended inspection. Calculations similar to the Figure 8.3 procedure were then made and showed that the backflow through the impeller wear ring clearance to the impeller inlet (eye) was easily as high as 30% of the BCP's rated flow capacity.

This surely explained why these pumps were never quite delivering their rated capacity, even just after a rebuild. From consultation with the pump OEM's lead hydraulics engineer, it was revealed, and not surprising, that such a high rate of wear ring back flow would significantly disrupt the impeller inlet flow velocity distribution, with the likely outcome of a quite high dynamic hydraulic impeller force at the vane passing frequency. The root cause smoking gun was finally found. The impeller wear ring needs to be replaced when the clearance wears open to twice the as-new value, an industry widely held criteria for centrifugal pumps.

8.5 Root Cause and Fixes

The excessive 6N vane-passing shaft vibration caused by enormous impeller wear ring backflow was the root cause. The fix is simply that the impeller wear ring must be replaced when the clearance wears open to twice the maximum as-new value. Since inspection of the wear ring is difficult, as explained in Section 8.4, the wear ring should be replaced regularly at times when the boiler is shut down long enough to pull the BCPs up to safely access the vacated impeller space and replace the wear ring. The pump OEM has available a zero leakage mechanical seal retrofit that eliminates the floating ring seals and thus eliminates the need for seal injection flow. However, even with the superior OEM seal retrofit mechanical seal, the wear ring clearance replacement criteria should still be followed.

9

Nuclear Plant Cooling Tower Circulating Pump

9.1 Problem Background

This case study deals with a nuclear power plant boiling water reactor (BWR) generating unit rated at 1300 MW. It has three cooling tower circulation pumps, requiring at full load only two of these pumps operating during the winter season, but all three operating during the summer season. These are quite large vertical rotational centerline pumps with a rotor speed of 325 rpm (5.5 Hz). Over a period of several years, at least one of these three pumps moved sufficiently out of plumb to need replumbing. That was accomplished during the winter season when this pump was taken out of service. Following this effort the unit was test run and found to have excessive levels of vibration at a rotor speed frequency of 5.5 Hz. It was suggested that the plumb shimming used unfortunately reduced floor contact, called "soft foot." A preliminary assessment was that as a result of the soft foot, a structural resonance frequency moved right down into the spin frequency.

9.2 Investigation and Root Cause

The author's team installed accelerometers at various elevations on the unit and confirmed that the excessive vibration was a resonance, based on acceleration measurements taken on a coast-down run. Figure 9.1 shows a sample of these acceleration measurements, displaying both the peak and a rapid phase angle change fundamentally characteristic of a resonance. Figure 9.2 illustrates the elevation view of the orbital vibration trajectory of the unit.

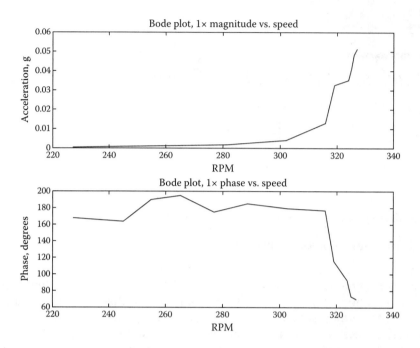

FIGURE 9.1
Measurements of cooling tower pump-motor centerline vibration.

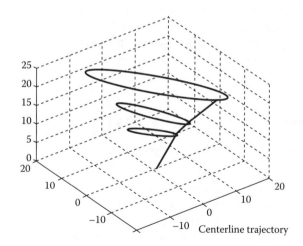

FIGURE 9.2
Orbital vibration trajectory of cooling tower pump-motor unit.

9.3 Fixes

The ultimate long-term fix was to eliminate the soft foot. A cost-effective intermediate fix recommended to the plant owner was to design and install a tuned vibration absorber attached to the top of the motor. The author and his staff designed a vibration absorber for this application. A tuned vibration absorber replaces the preexisting resonance with two side-band resonances, one below and one above, the preexisting resonance frequency. The absorber is simply an attached Spring (k) – Mass (m) system tuned to the preexisting natural frequency, that is, $\omega_n = \sqrt{k/m}$. The greater the absorber mass, the greater the frequency separation between the two resulting side-band resonance peak replacements. This is a common low-cost and effective fix in the field. Figure 9.3 shows absorber design analysis results for the final absorber design. The recommended absorber design is illustrated in Figure 9.4.

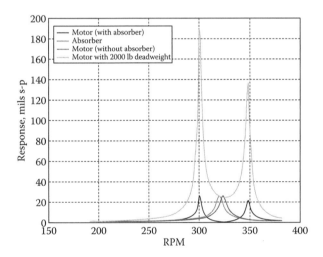

FIGURE 9.3
Computed vibration amplitude of pump motor.

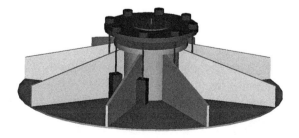

FIGURE 9.4
(See color insert.) Vibration absorber (red and blue) atop pump motor.

10

Condensate Booster Pump Shaft Bending

10.1 Problem Background

This condensate booster pump (CBP) is one of two 50% CBPs on a 600 MW unit. Both CBPs are required for full load operation, but one CBP can handle the required condensate flow at part load operation up to approximately 450 MW. The 1000 HP, 3580 rpm induction drive motor of the CBP had a several-month history of excessively high vibration levels. It was observed that this CBP's vibration levels go from bad to worse as the unit is operated from a cold start to steady-state operating temperatures, taking approximately 2 hours. All corrective actions undertaken by the plant to reduce of the motor's vibration levels to within acceptable amplitudes were unsuccessful. In the meantime, the operating vibration levels of this motor have slowly worsened, indicating an absolute need to solve the problem.

10.2 Investigation

A vibration specialist from the CBP's OEM took vibration measurements on both CBPs. Those measurements and corresponding analyses eliminated both base-motion characteristics as well as the pump internals as potential sources of the CBP's excessive motor vibration problem. This investigation combined with earlier observations that the CBP's vibration levels go from bad to worse as the unit is operated from a cold start to steady-state operating temperatures led the OEM's vibration specialist to conclude that the root cause of the vibration problem was thermal bowing of the motor rotor. This diagnosis would likely mean that local electrical shorting on the motor rotor windings produced local heating that led to a bow being imposed upon the motor shaft by the rotor windings. Based on discussions with the plant's local motor repair vendor, it was concluded that the most cost-effective solution for such a motor condition is replacement of the motor, not repair. OEM's vibration specialist recommended that measurement of the motor's shaft orbital displacement vibration during the thermal transient from a cold start to steady operating temperatures could clearly establish if the *bowed-rotor diagnosis* is correct. Accordingly, the author's team was retained to make motor vibration measurements, using accelerometers and X-Y displacement proximity probes. Those measurements and corresponding evaluations clearly confirmed the OEM's bowed-motor-rotor diagnosis. Figure 10.1 shows the vibration measurement points.

The vibration signals were recorded simultaneously in 10-second snippets taken at 10-minute intervals over the full-load 2-hour motor warm-up. Prior to the cold start-up and 2-hour continuous warm-up to steady temperatures, the unit was started for a few seconds and then shut down, so that proximity probe signals could be recorded for that coast down. That coast-down data at low speed (≈300 rpm) was used to subtract proximity-probe electrical runout from the total signals recorded subsequently during extended operation.

From reduced 300 rpm data, a 2.6 mil peak-to-peak shaft runout was found and was unexpectedly quite high for electrical runout alone, suggesting a bent motor shaft. A dial indicator was therefore subsequently set up to obtain cold shaft mechanical runout where the proximity probes had been located. By hand rotating the shaft in 45° increments, mechanical runout data was taken from the dial indicator and is compared to the coast down 300 rpm proximity probe runout data in Figure 10.2.

The comparison in Figure 10.2 is striking. It shows that the low-speed runout captured by the proximity probes is almost entirely due to actual mechanical runout, not proximity-probe sensed purely electrical runout. The actual electrical runout was thereby determined to be comparatively quite small, reflecting a pretest emery cloth and solvent cleaning of the proximity

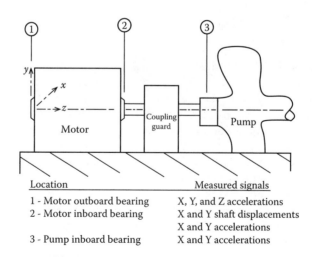

FIGURE 10.1
BCP pump-motor vibration measurement points.

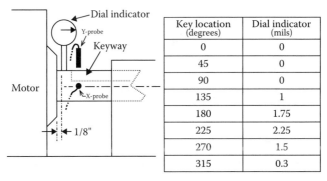

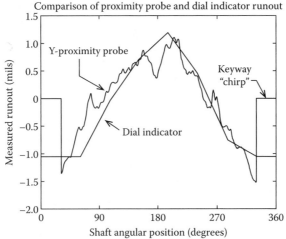

FIGURE 10.2
Motor shaft runout measurements.

probe targeted portion of the shaft. Realizing that the X-Y proximity probes and subsequent dial indicator were located axially close to where the bent shaft would contact the journal bearing edge at shut down (i.e., a runout nodal point), this measured mechanical runout indicated a seriously bent shaft, probably well in excess of 5 mils. This bent condition of the motor shaft was apparently the consequence of considerable thermal bowing as the motor heats up to operating temperature. A large thermal bow apparently locked in a shaft distortion by nonuniformly shifting the axial distribution of the heavy shrink-fit between the motor rotor and its shaft. It was also possible that some shaft yielding caused by the thermal bow could have also contributed to the permanent bent condition of the motor shaft.

The orbital displacement proximity probe data presented in Figure 10.3 reveals that the motor inboard shaft total orbital displacement vibration grew to 6 mils peak-to-peak at 26 minutes after the cold start, about double the value at the beginning of the test time. The 2-hour trend from cold start

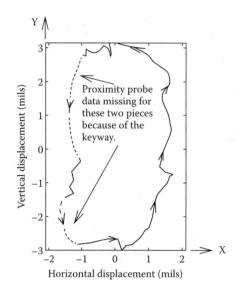

FIGURE 10.3
Shaft X-Y orbital vibration trajectory after warm-up.

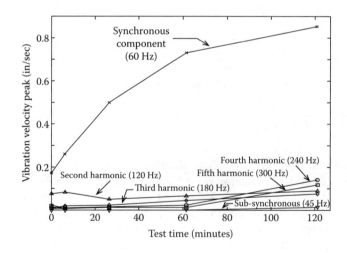

FIGURE 10.4
Motor shaft inboard vertical vibration velocity peaks for 2-hour test.

shown in Figure 10.4 typified the motor vibration signals in general, thus indicating thermal bowing of the motor shaft. As the motor heated up to operating temperature, the vibration level grew from rough to very rough, leveling out at nearly 1 in/sec peak vibration velocity. Figure 10.4 shows that it is primarily the synchronous (once-per-rev) vibration component that dominates the overall vibration levels on the motor. The outboard axial motor vibration (Z-direction) 2-hour trend also exhibited a similar strong growth of the synchronous vibration component to about 1 in/sec. Such a high once-per-rev axial vibration is a typical symptom of a bent shaft and/or badly bowed shaft (Adams 2010).

10.3 Root Cause and Fix

To a high degree of certainty, the root cause of this CBP motor vibration problem originated from a thermal bowing of the motor shaft, apparently produced by highly localized heating at a motor rotor winding short. Furthermore, the motor shaft was seriously bent as detected both with a dial indicator by hand rotating the shaft and from low-speed coast-down proximity probe signals. To a high degree of certainty, thermal bowing had caused this static bend in the shaft to be developed from a distorting axial redistribution of the heavy motor rotor-to-shaft shrink fit. Localized shaft material yielding may also have contributed to the shaft static bend as a result of the thermal bowing. Given the condition of the CBP motor rotor and shaft, replacement of the motor was surely the most cost-effective solution to the problem on this unit.

11

Pressurized Water Reactor Primary Coolant and Auxiliary Feed Pumps

11.1 Primary Coolant Pump (PCP) Problem Background

Chapter 5, Section 5.1 describes pressurized water reactor (PWR) primary coolant pumps (PCPs). Figures 5.4b and 5.6 show illustrations of a PCP. As detailed in Section 5.1, given the three-bearing rigidly coupled configuration typical of PCPs, unless the three journal bearings are perfectly aligned on a straight line, there will be additional journal-bearing static loads from the three bearings' inadvertent preloading of each other. The statically indeterminate journal bearing loads combined with the vertical rotational centerline (i.e., no rotor-weight biasing of radial loads) makes the journal bearing loads randomly variable. That of course makes PCP rotor dynamic characteristics randomly variable as well. Jenkins (1993) attests to the considerable challenge in assessing the significance of monitored vibration signals from PCPs, focusing upon the possible correlation of vibration signal content and equipment malfunction as related to machine age. Jenkins presents the *Westinghouse approach* in identifying vibration problem root causes and corrective changes for these aging PCPs.

11.2 Vibration Instrumentation False Alarm

For rotor vibration monitoring of PCPs, an X–Y pair of proximity probes are mounted 90° apart just below the coupling, targetting the short straight low-run-out section of the shaft. In one of Jenkins's case studies, what appeared to be a sudden unfavorable change in monitored rotor vibration orbits was in fact eventually traced to a combined malfunction and faulty installation of the eddy current proximity probes. This discovery also led to the finding that eddy current probe displacement systems are vulnerable to deterioration over time in the hot and radioactive environment around PCPs, and thus need to be replaced or at least checked at regular intervals. This vibration false-alarm case also emphasized the importance of closely following the proximity probe vibration instrumentation manufacturer's instructions regarding permissible part number with allowable shaft-target material combinations to avoid errors in probe-to-target scale factors upon which all inductance proximity probe signals are based.

11.3 Worn Impeller Journal Bearing

A second case study presented by Jenkins (1993) pertains to a PCP of the configuration illustrated in Figure 5.6 (see Chapter 5). It is one of the three identical pumps for a specific reactor. It developed a large half-frequency ($N/2$) rotor vibration whirl. With the high ambient pressure, thus cavitation-free water-lubricated impeller sleeve bearing on a vertical centerline, there is nearly always some tolerable $N/2$ vibration content observed in the monitored rotor vibration signals of these pumps. In this case, the drastically increased level of $N/2$ vibration led to an investigation to determine the likely root cause(s) and the proper corrective action(s). Based on both the drastic increase in monitored $N/2$ vibration component (changed from 2 to 6 mils p.p. at coupling) and on a shift in static centerline position as extracted from the proximity-probe DC voltages, it was diagnosed that the impeller journal-bearing clearance had significantly worn open. It was thus decided that replacement of the impeller bearing be replaced at least at the next refueling outage, or sooner if the monitored vibration developed a subsequent upward trend.

11.4 Primary Coolant Pump Summary

PCPs are not the only vertical pump used in power plant application. Some fossil-fired steam boilers in electric power-generating plants are designed with boiler circulating pumps (Chapter 8), which are incorporated into the design to make the boiler physical size much smaller than it would otherwise have to be if relying on free convection alone. Like PCPs, boiler circulating pumps have vertical centerlines dictated by suction and discharge piping constraints. Condensate pumps are another example of vertical centerline machines (Chapter 4, Section 4.2). Marscher (1986) presents a comprehensive experience-based treatment to vibration problems in these vertical pumps. Most large hydroelectric turbines and pump turbines are vertical.

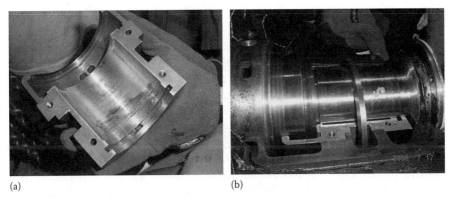

FIGURE 11.1
Auxiliary feed pump distressed: (a) bearing and (b) shaft.

11.5 Auxiliary Feed Pump Problem Background

The author's team was retained to participate in troubleshooting and analysis of a distressed auxiliary feed pump (AUXFP) oil-ring lubricated inboard journal bearing. Photos of the distressed bearing half and matting shaft are shown in Figure 11.1.

The available information for the trouble shooting and analysis included the distressed bearing and mating journal, maintenance and installation history, bearing temperature data, oil samples, and anecdotal experience from plant personnel. After independently analyzing the available information, a list of likely root causes was generated and corresponding recommendations were provided.

An inspection of the damaged bearings indicated with a high degree of certainty that bearing overloads were experienced on the pump side of the inboard bearing, and that these overloads caused the damage to the journal bearing babbitt surface. In light of the available information, the most likely root causes of the overload and resulting bearing damage are explained in the following section, in order of priority.

11.6 Auxiliary Feed Pump Analysis Results and Recommendations

Bearing retaining set screws were not installed with the OEM-recommended torque. In split-casing designs, it is common for designers to include a small but tightly controlled interference fit between the bearing and the bearing housing called "bearing pinch." Just enough bearing pinch is needed to keep the bearing in place and provide a very stiff support for the bearing that then also provides its available fluid-film rotor-dynamic stiffness and damping capability. But there should not be enough pinch to significantly distort the bearing shells. The design of the bearings is to achieve a similar pinch function through the use of two set screws, which thread through the bearing housing from the top half and secure the bearing into place. Too much torque on these set screws will distort the bearing, but too little will allow the bearing to move within the bearing housing in ways it was not intended to move. Such movement can cause the bearing to seat improperly in the bearing housing leading to localized misalignment and overloading, as well as excessive rotor vibration (see Chapter 6, Section 6.1). Care must be taken in setting the torque on both sets of motor bearing retaining set screws to the OEM's recommended torque values.

Internal misalignment. The machined clearance between the AUXFP bearing and the bearing housing is of the order of 0.002 to 0.006 inches. In the original design this amount of space may be required in order to account for a stack-up of tolerances or some other assembly/dimensional variability. But as far as the function of the bearings is concerned, this is considered a very loose fit and more than needed to accommodate differential thermal expansions. Furthermore, the dimensions for the bearings and bearing housing, as originally designed, appear to be specified with two decimal places, which generally correspond to a tolerance of ±0.010 inches. The possible stack-up of the tolerances combined with the general looseness of the fit has the potential to lead to difficulties in achieving a good internal alignment. If excessive internal misalignment exists, an improper torque on the bearing-retaining set screws may worsen the problem. It was recommended to check the internal alignment of the bearing housings without the bearings installed, and comparing that to the actual dimensions of the bearings to see if there is any potential stack-up of tolerances that should be accounted for during bearing installation.

Misalignment between the pump and motor caused by thermal growth. The pump and motor elevations should be set such that at the steady-state operating condition, the two shafts are aligned. This involves setting the motor shaft elevation slightly lower than the pump shaft elevation. Setting the motor too high may cause static load to be transferred through the coupling. On the other hand, if the coupling is functioning properly and has been designed to accommodate the expected thermal growth, it should not be transmitting any static radial loads and should not be a contributing factor. It was recommended that the motor-to-pump coupling should be checked for wear, and the allowable radial misalignment should be compared to the expected thermal growth values to make sure the coupling is appropriate for this application. Finally, it was suggested that the alignment procedure for this AUXFP be reviewed with the goal of more tightly controlling the alignment process for this pump.

The team reviewed the proposed course of action and subsequently witnessed the successful start-up and quarterly test of this unit. It is important to keep in mind that babbitted bearing surfaces are extremely forgiving, so it is quite probable (but difficult to definitively quantify) that such AUXFPs could run for an extended period of time even with such bearing surface damage, as encountered in this case.

12

Cases from Mechanical Solutions, Inc.

The following troubleshooting case studies have been provided for publication by Mechanical Solutions, Inc. (MSI) of Whippany, New Jersey (www.mechsol.com).

12.1 Below-Ground Resonance of Vertical Turbine Pump

Two vertical turbine pumps (VTPs) with 45 ft (13.72 m) long column assemblies (see Figure 4.3) installed in a nuclear facility were suffering from premature wearing of the line shaft bearings and shaft sleeves leading to frequent and costly repairs. The MSI measurement spectra plots (in logarithmic scale) shown in Figure 12.1 led to a preliminary conclusion that the root cause was a below-ground classic reed frequency problem. An accelerometer mounted on the suction bell flange (Figure 12.1b) shows that the third below-ground mode (15.9 Hz) is only 7% away from the 1x running speed (14.9 Hz, 894 rpm) excitation source. In order to design a solution and avoid costly and time-consuming trial-and-error problem solving, MSI performed an operating deflection shape (ODS) test. The ODS test results were used to calibrate a finite element analysis (FEA) model. The FEA model was used to determine potential solutions.

Figure 12.2 shows a freeze frame of an ODS animation showing the third below-ground mode at 15.9 Hz. A properly executed ODS exaggerated-displacement to-scale animation is a quite insightful way to understand and explain a structure's dynamic motion. The problematic below-ground third bending is clearly shown. One solution option was to add a number of 1-inch thick ribs to the upper column, shifting the third bending mode upward by 16% (well away from running speed). The second option was to shorten the column pipe, to shift the offending mode's natural frequency even higher. The plant elected to shorten the column after it was determined to be hydraulically acceptable.

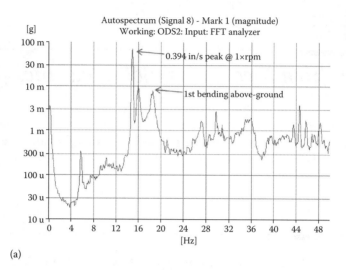

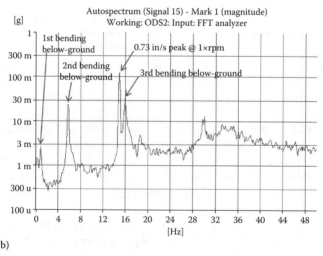

FIGURE 12.1
Vibration measurements on vertical turbine pump. (a) Measured on a motor bearing and (b) measured on suction bell flange.

12.2 Cavitation Surge in Large Double-Suction Pumps

As summarized in Section 3.5.4, a long-held consensus among pump hydraulic specialists is that the cavitation phenomenon does not lend itself to a reliable cavitation-damage predictive tool based upon measureable noise. So the challenge in this case was to apply a new nonintrusive test method to quantify the severity and rate of damage caused by cavitation. MSI was retained

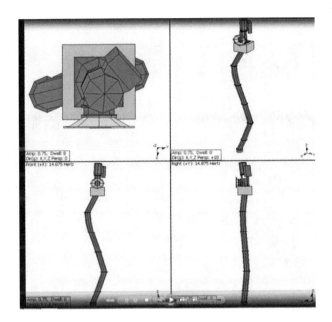

FIGURE 12.2
Animation freeze frame of the third below-ground mode at 15.9 Hz.

to apply its new high-frequency technique using external accelerometers to quantify damage occurring in a pump operating with significant cavitation. This new method has been used to diagnose cavitation in pumps, valves, and other fluid system components.

Large, recently installed double-suction pumps exhibited excessive auditory noise. MSI was retained to confirm if the pumps were causing the excessive noise and to determine whether the noise was a result of cavitation that could significantly reduce pump component life. Of particular concern was a periodic chugging sound that was present when the pumps were operating within a flow range of approximately 100% to 130% of the best efficiency point (BEP). The Architecture Engineer (AE), while quite experienced, had never previously encountered a symptom quite like this. MSI had already used its new approach in other plants to estimate the severity of pump cavitation damage and to determine if component modifications are needed, or determine if the cavitation noise is simply a nuisance. Figure 12.3 shows acceleration and pressure measurement signals taken on one of these pumps.

Since the vibration spikes are considerably higher in amplitude when the cavitation bubbles collapse (see Chapter 2, Section 2.3.1) on a surface as opposed to collapsing in the free stream, a quantitative assessment of the potential for cavitation damage to the pump is thereby made. According to MSI, measuring the sound frequency spectrum, as others have done, involves a less direct measurement since the effects must be transmitted from the pump casing to

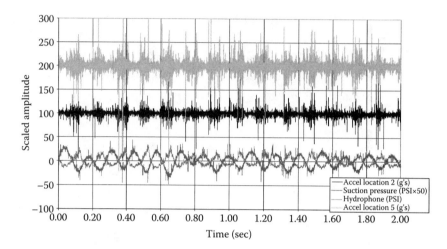

FIGURE 12.3
Acceleration, suction sound pressure, and pressure during cavitation.

the air before being measured. Even difficult-to-perform dynamic pressure or hydrophone measurements are less useful because they cannot determine whether the detected cavitation is harming internal metal surfaces. The tests run in this case consisted of measuring the pressure pulsations in the suction pipe, sound pressure near the impeller inlet, instantaneous casing acceleration at many surface locations, and sound pressure in the air surrounding the pumps. Based on past experience, the levels of peak instantaneous acceleration were indicative of damaging cavitation, 4 times above levels for which damage from cavitation would likely occur on either the casing fluid passages or on the impeller. Posttest inspection of pump internals revealed rapid cavitation erosion damage of the impeller vanes. In a similar pump investigation using MSI's technique at a major water plant, MSI had discovered that after 1000 operating hours at 3 times the acceleration limit, the 3/4-inch 316L impeller vanes had lost half their thickness at the cavitation-eroded locations.

Comparing the various measurement methods showed good correlation between suction pressure measurements, suction sound pressure measurements (hydrophone), and instantaneous acceleration. The measurement that provided the most ambiguous data was the more typically performed airborne sound pressure measurement. Discussions with the pump OEM revealed the pump suction had been oversized to meet stringent NPSH requirements at minimum suction head conditions. The result was a pump that experienced suction recirculation at flow rates even above BEP (typical recirculation cavitation damage occurs below BEP). The problem was compounded since the cavitation combined with suction recirculation set up an unsteady flow pattern at the pump suction, resulting in cyclic cavitation surging, not only damaging the impeller but also shaking the entire pump system. The problem was fully solved by an impeller redesign.

12.3 Impeller Vane-Pass Excitation of a Pipe Resonance in a Nuclear Plant

This case study concerns a chronic premature failure of bearings and seals in a safety-related *residual decay heat removal pump*, a single-stage end-suction volute pump (see Figure 5.6, RHR). The plant believed there was an inherent problem with the pump design. However, MSI's operating-deflection-shape testing demonstrated that the root cause for the component reliability problems was that the pump casing was being forced to distort dynamically, with a dominant frequency of impeller vane pass, driven by a vertical bouncing of the discharge pipe that jammed against the pump discharge nozzle. A time-averaged pulse modal test was performed and determined that a structural natural frequency dominated by vertical motion of the discharge pipe was nearly coincident with the first harmonic of the vane-pass frequency, thereby causing a resonance. This resonance was poorly damped, based on the half-power bandwidth (see Adams, 2010) of the resonant response peak for this mode determined from the modal test. Figure 12.4 shows the accelerometer locations.

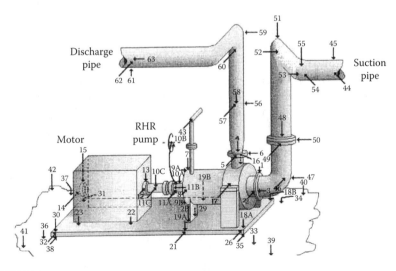

FIGURE 12.4
RHR pump assembly and modal test accelerometer locations.

MSI was requested by the regulatory agency and plant to verify that the structural resonance issue was the only reason for the increased vibration, and in particular that a piping system acoustic resonance was not in play. This was problematic, since there were no piping taps near the pump discharge, and if the piping was violated by installing a tap, it would need to be requalified for system operation in a safety-related situation. The cost of this requalification would be high, and the plant might be required to shut down until analysis was completed. MSI applied a technique that it had used successfully in other plants where pressure pulsation measurements were required in pump or compressor systems where many pressure readings at various locations along the pipe axis were needed to evaluate acoustic natural frequency mode shapes, but where pressure taps were sparse and/or impractical. The method consists of using a minimum of four uniaxial accelerometers, attached perpendicular to the pipe wall at 90° intervals around the periphery of the pipe at each location where pulsation readings are needed. Away from stiffening components such as flanges or piping supports, this technique allows the pressure pulsation (which makes the pipe radially expand) to be separated from the piping gross structural vibration, which merely translates the pipe as a relatively rigid body. These accelerometer measurements demonstrated that acoustics were not an issue.

MSI then used a test-calibrated finite element analysis (FEA) model to perform some what-if analyses of possible alternative solutions. Adding mass to the pipe, use of a piping damper (shock absorber), and stiffening of the pipe supports were all explored and all found to be viable solutions. However, each of these solutions would require piping requalification. MSI then designed a low weight vibration absorber (see cooling-tower pump example in Chapter 9, Figure 9.4), in the form of a low-mass thin-walled pipe clamshell surrounding the discharge pipe vertical leg, which is able to move vertically relative to the discharge pipe and attached to the discharge pipe through clamping plates that had tuned and adjustable stiffness in the vertical direction. This assembly was shown analytically and by test to quench the resonance at the vane-pass acoustic frequency. Because the assembly was of sufficiently low mass and did not penetrate the discharge pipe, only the vibration absorber needed analysis and review and regulatory approval.

Bibliography

Adams, M. L., Aging and low-flow degradation of auxiliary feedwater pumps, Proceedings of U.S. Nuclear Regulatory Commission, Aging Research Information Conference, March 1992.

Adams, M. L. (principal investigator), Model-based pump condition monitoring project, CWRU, 2016.

Adams, M. L., *Rotating Machinery Vibration: From Analysis to Troubleshooting*, 2nd ed., Taylor & Francis/CRC Press, 2010.

Adams, M. L., and Gates, W., Successful troubleshooting a nuclear feed water pump vibration problem, Proceedings of ASME-NRC Annual Conference, Washington, DC, June 2002.

Adams, M. L., and Loparo, K. A., Analysis of rolling element bearing faults in rotating machinery: Experiments, modeling, fault detection and diagnosis, Proceedings of IMechE 8th International Conference on Vibration and Rotating Machinery, Swansea, Wales, September 2004.

Adams, M. L., Loparo, K. A., Kadambi, K., and Zeng, D., Model-based condition monitoring for critical pumps in PWR and BWP nuclear power plants, Case School of Engineering Report to the U.S. Nuclear Regulatory Commission, 2004.

Buchter, H. H., *Industrial Sealing Technology*, Wiley, 1979.

Casada, D., and Adams, M. L., Low-flow testing of safety related nuclear pumps and the regulatory implications, Proceedings of EPRI Symposium on Power Plant Pumps, Tampa, Florida, June 1991.

Childs, D., *Turbo Rotordynamics: Phenomena, Modeling, and Analysis*, Wiley, New York, 1993.

Childs, D., and Vance, J. M., High pressure floating ring oil seals, Proceedings of Texas A & M Turbomachinery Symposium, 1997, pp. 201–220.

Dahlheimer, J. A., Elhauge, E. E., Greenberg, L., Jacobs, J. H., Kabbert, W. J., Keel, H. R., Masche, G. C., Moran, C. N., Petrie, D. H., Seid, R., and Weiss, T. G., *Westinghouse Pressurized Water Reactor Nuclear Power Plant*, Westinghouse Electric Corporation, 1984.

Dufour, J. W., and Nelson, W. E., *Centrifugal Pump Sourcebook*, McGraw-Hill, 1993.

Eshleman, R. L., Machinery Vibration Analysis II, Short Course, 1999.

Florjancic, D., *Development and Design Requirements for Modern Boiler Feed Pumps*, EPRI Symposium Proceedings: Power Plant Feed Pumps—The State of the Art, EPRI CS-3158, Chapter 4, pp. 74–99, July 1983.

Florjancic, D., *Troubleshooting Handbook for Centrifugal Pumps*, Turbo Institute, Ljubljana, Slovenia, 2008.

Guelich, J. F., and Bolleter, U., Pressure pulsations in centrifugal pumps, *ASME Journal of Acoustics and Vibrations* 114(2), 1992: 272–279.

Guelich, J. F., Bolleter, U., and Simon, A., Feed pump operation and design guidelines, EPRI Final Report TR-102102, Research Project 1884-10, 1993.

Guelich, J. F., Florjancic, D., and Pace, S. E., Influence of flow between impeller and casing on part load performance of centrifugal pumps, *Proceedings of 3rd ASCE/ASME Mechanical Conference*, San Diego, 1989, pp. 227–235.

Guelich, J. F., and Pace, S., Quantitative prediction of cavitation erosion in centrifugal pumps, Proceedings of International Association for Hydro-Environment (IAHR) Symposium, Montreal, 1986.

Hammit, F. G., *Cavitation and Multiphase Phenomena*, McGraw Hill, 1980.

Horattas, G. A., Adams, M. L., and Dimofte, F., Mechanical and electrical run-out removal on a precision rotor-vibration research spindle, *ASME Journal of Acoustics and Vibration* 119(2), 1997: 216–220.

Jenkins, L. S., Troubleshooting Westinghouse reactor coolant pump vibrations, EPRI Symposium on Trouble Shooting Power Plant Rotating Machinery Vibrations, San Diego, May 19–21, 1993.

Karassik, I. J., and Carter, R., *Centrifugal Pumps*, McGraw-Hill, 1960.

Karimi, A., and Avellan, F., Comparison of erosion mechanisms in different types of cavitation, *Wear* 113, 1986: 305–322.

Laberge, K., Shaft crack detection from axially propagating stress waves of cracked closings under rotation, PhD thesis, Case Western Reserve University, 2009.

Laberge, K., and Adams, M. L., Analysis of the elastic wave behavior in a cracked shaft, Proceedings of ASME, IDET/CIE, 2007.

Lebeck, A. O., *Principles and Design of Mechanical Face Seals*, Wiley, 1991.

Makay, E., How close are your feed pumps to instability-caused disaster, *Power Magazine*, 1980, pp. 69–71.

Makay, E., Private communication with M. L. Adams, 1977–1994.

Makay, E., Survey of feed pump outages, EPRI Research Project RP 641, Final Report FP-754, 1978 (edited by M. L. Adams).

Makay, E., and Adams, M. L., Operation and design evaluation of main coolant pumps for PWR and BWR service, EPRI Final Report NP-1194, 1979.

Makay, E., Adams, M. L., and Shapiro, W., *Design and Procurement Guide for Primary Coolant Pumps Used in Light-Water-Cooled Nuclear Reactors*, Oak Ridge National Laboratory, ORNL TM-3956, 1972.

Makay, E., and Barrett, J., Changes in hydraulic component geometries greatly increased power plant availability and reduced maintenance costs: Case histories, Proceedings of Texas A&M 1st International Pump Symposium, 1984.

Makay, E., and Szamody, O., *Recommended Design Guidelines for Feedwater Pumps in Large Power Generating Unit*, EPRI Report CS-1512, September 1980.

Marscher, W. D., The effect of fluid forces at various operating conditions on the vibrations of vertical turbine pumps, seminar by the Power Industries Division of IMechE, London, February 5, 1986.

Peterson, M. B., and Winer, W. O., *Wear Control Handbook*, ASME, 1980.

Pfleiderer, C., *Die Kreiselpumpen*, Springer-Verlag, 1932.

Robertson, M., and Baird, A., Thermodynamic pump performance monitoring in power stations, IMechE Seminar, Fluid Machinery in the Power Industry, Bristol, UK, June 10, 2015.

Stepanoff, A. J., *Centrifugal and Axial Flow Pumps*, 2nd ed., Wiley, 1957.

Suh, N. P., *The Delamination Theory of Wear*, Elsevier Sequoia S.A., Lausanne, 1977. (Also a series of papers published in *Wear* 44(1), 1977.)

Thoma, D., Veralten einer Kreiselpumpe bein Betrieb im Hohlzog Bereich, *Zeitschrift des Vereines Deutscher Ingenieure* 81, 1937: 972.

Index

Page numbers followed by f and t indicate figures and tables, respectively.

A

Abrasive wear, 64; *see also* Wear
Accelerometers, 29, 73, 74f
    CWRU three-stage test pump, 78, 80f
Adhesive wear, 62–64, 62f
Archard's law, 63, 64
Atomic energy, 95
Auxiliary feed pumps, 107
    analysis and recommendations, 145–147
    problem background, 144
Axial thrust bearing, 36, 39

B

Balance-disk leak-off flow, 76, 77f
Barrel-type boiler feed water pump, 20
Bearings
    damage, 70–71
    description, 35
    fluid-film, 35
    rolling-contact, 35–36, 36f
Below-ground resonance, of VTP, 149, 151f
Boiler circulation pump (BCP), 92f, 93
    components, 128f
    cross section, 128f
    floating-ring seal leakage, 130–131, 130f
    troubleshooting case studies, 127–132
    wear ring leakage, 131
Boiler feed water pumps, 85–88, 86f
    barrel-type, 20
    rotor sectional view, 114, 114f
    rotor unbalance and critical speeds, 113–122
Boiling water reactor (BWR)
    description, 96, 98f
    recirculation system, 108–109, 109f
    variable speed motors, 109
Boundary lubrication, 62

C

Campbell diagram, 54f, 55
Carbon-moderated air-cooled reactor, 95
Case Western Reserve University (CWRU), 78
    multistage centrifugal pump, 78–82, 78f–79f, 81f
Cavitation, 7
    damage caused by, 61
    description, 25–27
    in double-suction pumps, 150, 151–153
    laboratory shop testing, 27–28
    noise measurement, 78
Centrifugal pumps; *see also* Fossil plants; Nuclear power plant; *specific* pumps
    configurations, 16–22
    controls, 22
    discharge flow paths, 16–20
    entrance/inlet flow path, 16–20
    flow complexity and geometry, 3–10
    H-Q curves, *see* Head-capacity H-Q curve
    internal flows, 3
    intersection of, 24–25, 25f
    multistage, 7, 8f
    overview, 3
    performance and efficiencies, 23–24
    priming, 21, 22
    specific speed, 15–16
    suction nozzles, 6–7, 6f
    theory of, 10–16
    turbines *vs.*, 3–4, 4f
    volute suction nozzle, 7, 7f
Charging pumps, 108
Close-coupled vertical-centerline pumps, 30
Cold welding, 62

157

Combined-cycle plants, 20, 21
Condensate booster pump (CBP), 137–141
  orbital displacement proximity probe, 139–140, 140f
  shaft runout, 138–139, 139f
  vibration measurements, 137–138, 138f
Condensate pumps, 89–91
Condenser circulating pumps, 91, 92f
Constant-speed drives, 47
Controlled harmonic voltage, 69
Controls, centrifugal pumps, 22
Coolant, 95
Cooling tower circulation pumps, 133–135
  acceleration measurements, 133, 134f
  orbital vibration trajectory, 134f
  vibration absorber, 135, 135f
Coulomb friction coefficient, 63
Couplings, 30, 33–35
  diaphragm type, 35, 35f
  flexible, 30, 33–35
  flexing-element, 34, 34f
  gear, 33–34, 33f
  grid, 34, 34f
  rigid adjustable, 30, 32f
Critically damped condition, 57
Critical speeds, 55
  boiler feed pumps, 113–122
  natural frequencies, 55–57
Cylindrical journal bearing, 36, 37f

D

Degree-of-freedom (DOF) model, 54
  rotor vibration, 54–57
Delamination wear, 64
Diaphragm type coupling, 35, 35f
Diffuser, 3, 18
  flow separation, 5
  nozzle *vs.*, 4–5, 5f
  volute *vs.*, 18, 19f
Dimensionless numbers, 15
Discharge flow paths, 16–20
DOF, *see* Degree-of-freedom (DOF) model
Double-suction pumps, 7, 9f

  acceleration, 151, 152f
  auditory noise, 151
  cavitation in, 150, 151–153
  hydrophone measurements, 152
  pressure measurement, 151, 152f
Drive/drivers, 47–48
  constant-speed, 47
  variable speed, 48
Drooping *H-Q* curve, 49–50, 50f
Dry lubricant, 62
Dynamic seals, 39

E

Electric Power Research Institute (EPRI), 66

F

Face seals, 42, 43, 43f
  fluid-film variation, 43
Fast Fourier transform (FFT), 74, 76f
FEA, *see* Finite element analysis (FEA)
Feed water pump, 107
  cyclic thermal bow analysis, 125
  cyclic vibration symptom, 123
  double-suction, 7, 9f
  low-cost fix, 125–126
  radially split ring-section, 21, 21f
  rotor vibration analyses, 123
  shop cyclic thermal test, 125–126
  troubleshooting case study, 123–126
Finite element analysis (FEA), 149, 154
Fission, *see* Nuclear fission
Flexible couplings, 30, 33–35
  flexing-element, 34, 34f
  gear, 33–34, 33f
  grid, 34, 34f
Flexing-element couplings, 34, 34f
Floating-ring seals, 43, 44, 45f
Flow separation, 5
Fluid-film bearings, 35, 71
Fossil plants, 85–92
  BCP, 92f, 93; *see also* Boiler circulation pump (BCP)
  boiler feed water pumps, 85–88, 86f
  condensate pumps, 89–91, 89f

Index 159

condenser circulating pumps, 91, 92f
heater drain pumps, 85, 89–91, 90f, 91f
overview, 85
Fracture mechanics, 64
Fretting, 64
Fuel rods, 95

G

Gradually approached failures, 65
Grid coupling, 34, 34f

H

Harmonic current output, 69
Head-capacity H-Q curve, 25, 49–51, 50f
Heat, atomic energy converted into, 95
Heater drain pumps, 85, 89–91, 90f, 91f
Hertzian contact stress theory, 36
High pressure safety injection (HPSI) pump, 107–108
H-Q curve, *see* Head-capacity H-Q curve
Hydraulic efficiency, 23–24, 24f
Hydraulic forces, at off-design operating conditions, 66–69
Hydraulic instability, 49–51, 53, 70
Hydrodynamic axial-thrust bearing, 38f, 39
Hydrodynamic journal bearings, 36, 37f
Hydro turbines
  impellers, 3–4, 4f, 5
  vs. centrifugal pumps, 3–4, 4f, 5
Hysteresis loop, 56

I

Impellers, centrifugal pumps; *see also specific* pump
  efficiency, 5–6
  energy imparted to fluid by, 7, 8, 9–10
  flow entrance and exit, 10, 11f
  geometric configuration of, 6

inlet flow and discharge flow paths, 16–20
net driving power, 11–12
net through flow, 10
turbines *vs.*, 3–4, 4f, 5
velocity triangles, 10–15, 11f
Inlet flow
  paths, 16–20
  unfavorable conditions, 70
Intersection of centrifugal pumps, 24–25, 25f

L

Laboratory shop testing, 27–28
Lateral rotor vibration (LRV) analyses, 54
LOCA, *see* Loss-of-coolant accident (LOCA)
Lomakin effect, 40–42
Loss-of-coolant accident (LOCA), 107, 108
LRV, *see* Lateral rotor vibration (LRV) analyses

M

Massachusetts Institute of Technology (MIT), 64
Mechanical efficiency, 24
Mechanical losses, 24
Mechanical Solutions, Inc. (MSI) troubleshooting case studies, 149–154
Minimum recirculation flow, 52–53
Model-based conditioning monitoring, 78–82
Moderator, 95
Monitoring and diagnostics
  cavitation noise measurement, 78
  model-based, 78–82
  overview, 72
  pressure pulsation measurement, 76–77
  temperature measurements, 77–78
  vibration measurement, 73–76
Multiple-tongue volutes, 18
Multistage pumps, 7, 8f

external and internal return channels of, 20–21
recommended monitoring for, 82t

N

NASA, 66
Natural frequencies, 55–57
Net positive suction head (NPSH), 27, 28f, 61
   required and available, 28–29, 28f
Newton's second law for rotational motion, 11
Nozzle, 3
   diffuser *vs.*, 4–5, 5f
   flow separation, 5
NPSH, *see* Net positive suction head (NPSH)
Nuclear fission, 95, 96f
Nuclear power plant, 95
   auxiliary feed pumps, 107
   charging pumps, 108
   cooling tower circulation pumps, 133–135
   feed water pump, 107
   HPSI pump, 107–108
   major pumps, 104f
   RHRS, 107
   safety, recommended research areas, 104f
   standby safety pumps, 109–110
Nuclear reactor system, 95

O

Off-design operating conditions, hydraulic forces at, 66–69
Operating deflection shape (ODS) test, 149, 153
Operating failures
   excessive vibration, *see* Vibration
   hydraulic instability, *see* Hydraulic instability
   minimum recirculation flow, 52–53
   overview, 49
   pressure pulsations, *see* Pressure pulsations
   problem modes, 64–72
   wear, 60–64

P

Packing gland, 42
Packing shaft seal, 42, 42f
Parallel operation, 29, 29f
Pivoted pad journal bearings (PPJB), 36, 37f–38f
Pressure pulsations, 51–52, 52f, 68, 68f
   measurement, 76–77, 77f
Pressurized water reactor (PWR), 43, 44f, 98–106
   components, 95
   coolant loops for, 95, 98, 99f
   description, 95, 97f
   PCP, *see* Primary coolant pump (PCP)
   steam generator, 95, 102f
Primary coolant pump (PCP)
   aging, 106
   German design, 106, 106f
   journal bearing loads, 143
   sources of bearing loads, 105f
   vibration instrumentation false alarm, 143–144
   worn impeller journal bearing, 144
Priming, 21, 22
Pump casing, 6
   inlet flow and discharge flow paths, 16–20
Pump shafts, *see* Shafts

R

Raceway spalling, 64
Radial-contact ball bearings, 36, 39
Radially split ring-section, 21, 21f
RCS, *see* Reactor coolant system (RCS)
Reactor coolant system (RCS), 107
Recirculation system, BWR, 108–109, 109f
Residual decay heat removal (RHR) pump, 107
   accelerometer locations, 153f
   ODS testing, 153
   time-averaged pulse modal test, 153
   vane-pass frequency of resonance, 153–154
Residual heat removal system (RHRS), 107

Resonance-vibration-amplitude, 56–57
Reynolds lubrication equation, 36
RHRS, *see* Residual heat removal system (RHRS)
Rolling-contact bearings, 35–36, 36f, 70–71
Rotating-machine shafts, 30
Rotating shaft seals, 39, 40f
Rotational motion, Newton's second law for, 11
Rotor
   dynamic forces acting on, 59, 60f
   model, 54f, 55
Rotor dynamic characteristics, 66
Rotor mass unbalance, 55, 65–66
   boiler feed water pumps, 113–122
Rotor vibrations
   DOF model, 54–57
   self-excited, 57–58

S

Saddle-shaped *H-Q* curve, 50, 50f
Safety injection system (SIS), 107–108
Seals, 39–44
   categories/types, 39f
Self-priming, 21, 22
Shaft breakage, 71
Shafts, 30
   damage, 71
   failure, 30, 32f
   seals, 39–44; *see also* Seals
Shallow groove configuration, 41f, 42
Single-plane in-service balancing shots, 113
Single-tongue volutes, 17–18, 17f
SIS, *see* Safety injection system (SIS)
Smooth bore bushings, 39–40
Source rods, 95
Spalling, 36
Specific speed, 15–16
Speed
   critical, *see* Critical speeds
   specific, 15–16
Standby safety pumps
   quarterly testing of, 109–110
   stressor, 110
Static bending sag line, 71
Static equilibrium, 63

Static seals, 39
Steam generator, 95, 102f
Steam turbine-generator
   FEA rotor model, 54f, 55
   for nuclear plant, 96, 103f
Stepped-land seal, 43
Subsurface initiated high-cycle material fatigue, 35
Suction nozzles, 6–7, 6f
Sudden failures, 65

T

Tapered-land fluid film seal, 43
Temperature measurements, 77–78
Theory, of centrifugal pumps, 10–16
Thrust balancers, 45–47
   damage, 71–72
Two-tongue volutes, 17f, 18

U

Underdamped systems, 57
United States, 95
Uranium (U-235), 95
U.S. Nuclear Regulatory Commission, 96

V

Variable speed drives, 48
Variable speed motors, 109
Velocity severity levels, 73–74
Velocity triangles, 10–15, 11f
Vertical turbine pumps (VTP)
   below-ground resonance of, 149, 151f
   vibration measurements on, 149, 150f
Vibration, 53–59; *see also* Operating failures
   excessive, 53–59
   measurement, 73–76
   self-excited rotor, 57–58
Vibration displacement severity criteria, 74, 75t
Volutes, 17–19
   diffuser *vs.*, 18, 19f
   multiple-tongue, 18
   single-tongue, 17–18, 17f
   two-tongue, 17f, 18

W

Water-cooled reactors, 95; *see also* Boiling water reactor (BWR); Pressurized water reactor (PWR)
Wear, 60–64
    abrasive, 64
    adhesive, 62–64
    cavitation, 61
    delamination, 64
Wear Control Handbook (Peterson and Winer), 60
Wear rings, 39
Westinghouse approach, 143